Make:

Forrest Mims' Science Experiments

DIY Projects from the Pages of Make:

Forrest Mims

Forrest Mims' Science Experiments
DIY Projects from the Pages of Make:
By Forrest Mims

Printed in Canada.

Published by Maker Media, Inc.,
1160 Battery Street East, Suite 125
San Francisco, California 94111

Maker Media books may be purchased for educational, business, or sales promotional use. Online editions are also available for most titles (safaribooksonline.com). For more information, contact our corporate/institutional sales department: 800-998-9938 or corporate@oreilly.com.

Publisher: Roger Stewart
Editor: Patrick DiJusto
Copy Editor: Nancy Peterson, Happenstance Type-O-Rama
Proofreader: Scout Festa, Happenstance Type-O-Rama
Interior Designer and Compositor: Maureen Forys, Happenstance Type-O-Rama
Illustration: Richard Sheppard, Happenstance Type-O-Rama
Cover Designer: Maureen Forys, Happenstance Type-O-Rama
Indexer: Valerie Perry, Happenstance Type-O-Rama

September 2016: First Edition

Revision History for the First Edition
2016-07-25: First Release

See oreilly.com/catalog/errata.csp?isbn=9781680451177 for release details.

978-1-680-45117-7 [2016-09-23]

Safari® Books Online
Safari Books Online is an on-demand digital library that delivers expert content in both book and video form from the world's leading authors in technology and business.

Technology professionals, software developers, web designers, and business and creative professionals use Safari Books Online as their primary resource for research, problem-solving, learning, and certification training.

Safari Books Online offers a range of plans and pricing for enterprise, government, education, and individuals. Members have access to thousands of books, training videos, and prepublication manuscripts in one fully searchable database from publishers like O'Reilly Media, Prentice Hall Professional, Addison-Wesley Professional, Microsoft Press, Sams, Que, Peachpit Press, Focal Press, Cisco Press, John Wiley & Sons, Syngress, Morgan Kaufmann, IBM Redbooks, Packt, Adobe Press, FT Press, Apress, Manning, New Riders, McGraw-Hill, Jones & Bartlett, Course Technology, and hundreds more. For more information about Safari Books Online, please visit us online.

How to Contact Us
Please address comments and questions concerning this book to the publisher:

Make:
1160 Battery Street East, Suite 125
San Francisco, CA 94111
877-306-6253 (in the United States or Canada)
707-639-1355 (international or local)

Make: unites, inspires, informs, and entertains a growing community of resourceful people who undertake amazing projects in their backyards, basements, and garages. Make: celebrates your right to tweak, hack, and bend any technology to your will. The Make: audience continues to be a growing culture and community that believes in bettering ourselves, our environment, our educational system—our entire world. This is much more than an audience; it's a worldwide movement that Make: is leading and we call it the Maker Movement.

For more information about Make:, visit us online:

- Make: magazine makezine.com/magazine
- Maker Faire makerfaire.com
- Makezine.com makezine.com
- Maker Shed makershed.com
- To comment or ask technical questions about this book, send email to bookquestions@oreilly.com.

Dedication

This book is dedicated to my late father, Forrest M. Mims Jr., and my wife Minnie, both of whom encouraged my pursuit of science, and our three children, Eric, Vicki and Sarah, each of whom produced outstanding science fair projects during their school years.

Acknowledgments

This book owes much to former *MAKE:* Magazine editor Mark Frauenfelder, who understands better than anyone the motivations that drive and inspire both makers and amateur scientists. It was Mark who assigned the column in *MAKE:* that evolved into this book.

Contents

Preface:

Becoming an Amateur Scientist

An editorial in a leading science journal once proclaimed an end to amateur science: "Modern science can no longer be done by gifted amateurs with a magnifying glass, copper wires, and jars filled with alcohol." I grinned as I read these words. For then as now there's a 10× magnifier in my pocket, spools of copper wire on my workbench, and a nearby jar of methanol for cleaning the ultraviolet filters in my homemade solar ultraviolet and ozone spectroradiometers. Yes, modern science uses considerably more sophisticated methods and instruments than in the past. And so do we amateurs. When we cannot afford the newest scientific instrument, we wait to buy it on the surplus market or we build our own. Sometimes the capabilities of our homemade instruments rival or even exceed those of their professional counterparts.

So began an essay about amateur science I was asked to write for *Science* (April 1999, http://science.sciencemag.org/content/284/5411/55.full), one of the world's leading science journals. Ironically, the quotation in the first sentence came from an editorial that *Science* had previously published.

In the years since my essay appeared in *Science*, amateur scientists have continued doing what they've done for centuries. They've discovered significant dinosaur fossils, found new species of plants, and identified many new comets and asteroids. Their discoveries have been published in scientific journals and books. Likewise, thousands of websites detail an enormous variety of amateur science tips, projects, activities, and discoveries. Ralph Coppola has listed many of these sites in "Wanderings," his monthly column in *The Citizen Scientist* (www.wanderings.ca/TNW/Archive/Wanderings.pdf).

Today's amateur scientists have access to sophisticated components, instruments, computers, and software that could not even be imagined back in 1962 when I built my first computer, a primitive analog device that could translate 20 words of Russian into English with the help of a memory composed of 20 trimmer resistors (www.digibarn.com/stories/MITS/forrest-mims-III-material/Homebrew%20Analog%20Computer.pdf).

Components like multiwavelength LEDs and laser diodes can be used to make spectroradiometers and instruments that measure the transmission of light through the atmosphere. Images produced by digital video and still cameras can be analyzed with free software like ImageJ to study the natural world in ways that weren't even imagined a few decades ago. Amateur astronomers can mount affordable digital cameras on their telescopes, which then scan the heavens under computer control.

Cameras, microscopes, telescopes, and many other preassembled products can be modified or otherwise hacked to provide specialized scientific instruments. For example, digital camera sensors are highly sensitive to the near-infrared wavelengths beyond the limits of human vision from around 800nm–900nm. IR-blocking filters placed over camera sensors block the near-IR so that photographs depict images as they'd be seen by the human eye. Removing the near-IR filter provides a camera that can record the invisible wavelengths reflected so well by healthy foliage.

Many of the makers who publish their projects in the pages of *MAKE*, *Nuts and Volts*, and across the web have the technical skills and resources to devise scientific tools and instruments far more advanced than anything my generation of amateur scientists designed. They also have the ability to use these tools to begin their own scientific measurements, studies, and surveys. Thus, they have the potential to become the pioneers for the next generation of serious amateur scientists.

This book covers some of the many ways you can enter the world of amateur science. For now I'll end this chapter with a brief account of how I began doing serious amateur science, so you can see how a relatively basic set of observations of the atmosphere has lasted more than 20 years and, with any luck, will continue for another 20 years.

Case Study: 20 Years of Monitoring the Ozone Layer

In May 1988, I read that the U.S. government planned to end a solar ultraviolet-B radiation-monitoring program due to problems with instruments. Within a few months I began daily UVB monitoring using a homemade radiometer. The radiometer used an inexpensive op-amp integrated circuit to amplify the current produced by a UV-sensitive photodiode. An interference filter passed only the UVB wavelengths from about 300nm–310nm, while blocking the visible wavelengths.

I described how to make two versions of the UVB radiometer in "The Amateur Scientist" column in the August 1990 *Scientific American*. This article also described how the radiometer detected significant reductions in solar UVB when thick smoke from forest fires at Yellowstone National Park drifted over my place in South Texas in September 1988.

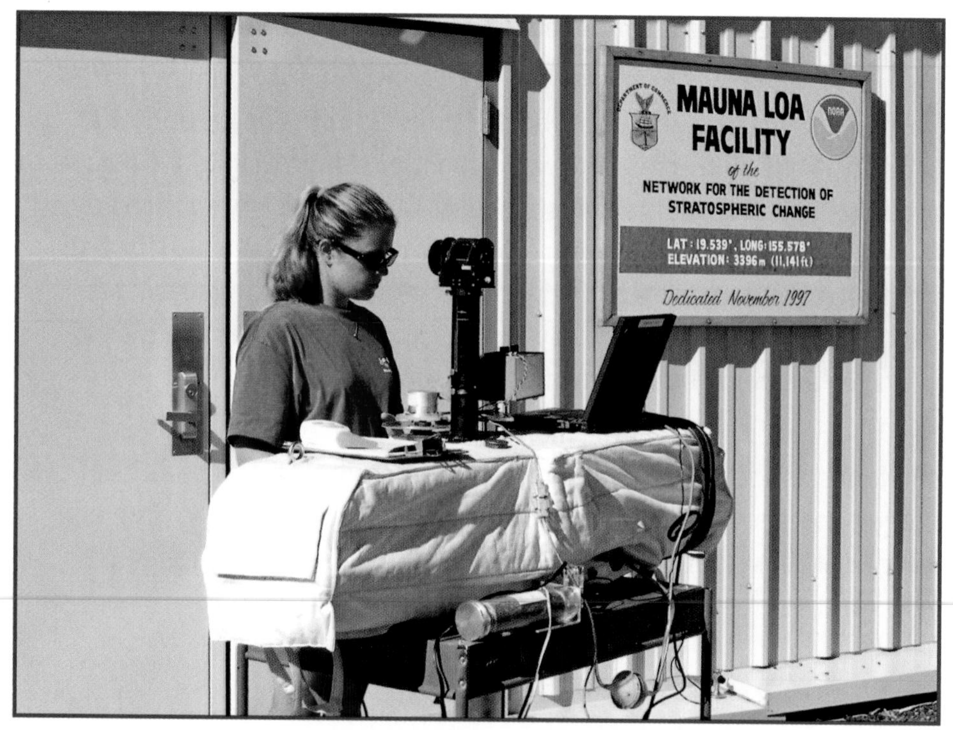

FIGURE P-1. Scientist Brooke Walsh measures the ozone layer with the world-standard ozone instrument at Hawaii's Mauna Loa Observatory, which Forrest Mims calibrated at the same location during summer 2016.

Ozone strongly absorbs UV, and you can determine the amount of ozone in a column through the entire atmosphere layer by comparing the amount of UV at two closely spaced UV wavelengths. This is possible because shorter wavelengths are absorbed more than longer wavelengths.

This meant that my simple UVB radiometer formed half of an ozone monitor. So I built two radiometers inside a case about half the size of a paperback book. One radiometer's photodiode was fitted with a filter that measured UVB at 300nm, and the second was fitted with a 305nm filter. I named the instrument TOPS for Total Ozone Portable Spectrometer. (Full details are at http://forrestmims.org/images/SCIENCE_PROBE_TOPS_PROJECT_NOV_1992_small.pdf.)

TOPS was calibrated against the ozone levels monitored by NASA's Nimbus-7 satellite. This provided an empirical algorithm that allowed TOPS to measure the ozone layer to within about 1% of the amount measured by the satellite. During 1990, ozone readings by TOPS and Nimbus-7 agreed closely. But in 1992, the two sets of data began to diverge so that TOPS was showing several percent more ozone than the satellite.

When I notified the ozone scientists at NASA's Goddard Space Flight Center (GSFC) about the discrepancy, they politely reminded me that the satellite instrument was part of a major scientific program and not a homemade instrument. I responded that I had built a second TOPS and both showed a similar difference, but this didn't convince them.

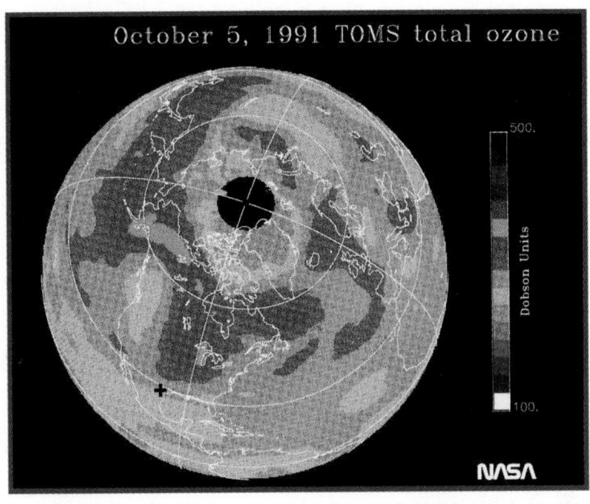

FIGURE P-2. This global ozone image was acquired while NASA's Nimbus-7 satellite was providing accurate data during 1991. On this day TOPS-1 measured 284.4 Dobson units (DU) of ozone, and the satellite measured 281.5 DU.

During August of 1992, I visited Hawaii's Mauna Loa Observatory for the first time to calibrate my instruments at that pristine site 11,200 feet above the Pacific Ocean. The world-standard ozone instrument was also being calibrated there, and it indicated a difference in ozone measurements made by Nimbus-7 that were similar to what I had observed.

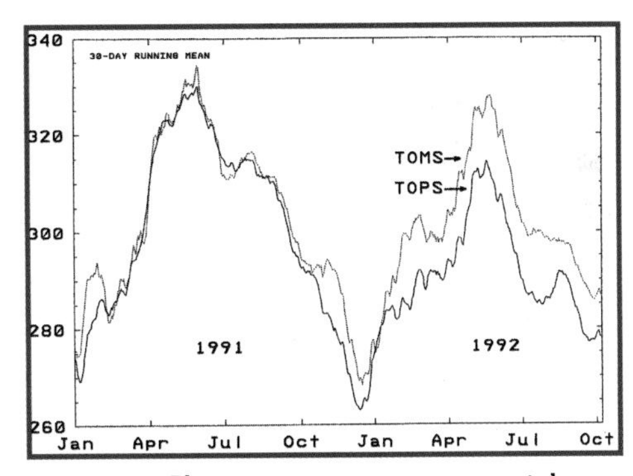

FIGURE P-3. Plot compares ozone measurements by TOPS and Nimbus-7. In 1992, the calibration of the satellite's instrument began to drift.

Eventually NASA announced that there was indeed a drift in the calibration of its satellite ozone instrument. A paper I wrote about this sparked my career as a serious amateur scientist when it was published in *Nature*, another leading science journal ("Satellite Ozone Monitoring Error," page 505, Feb. 11, 1993).

Later GSFC invited me to give a seminar on my atmospheric measurements titled "Doing Earth Science on a Shoestring Budget." That talk led to two GSFC-sponsored trips to study the smoky atmosphere over Brazil during that country's annual burning season, and several trips to major forest fires in the western U.S.

Going Further

The regular ozone measurements I began on Feb. 4, 1990 have continued to this day along with measurements made by various homemade instruments of the water vapor layer, haze, UVB, and other parameters. In future chapters we'll explore how you can also make such measurements—and make discoveries of your own.

FIGURE P-4. The TOPS project earned a 1993 Rolex Award that provided funds for the development of a first-generation micro-processor-controlled TOPS (Microtops) by Scott Hagerup.

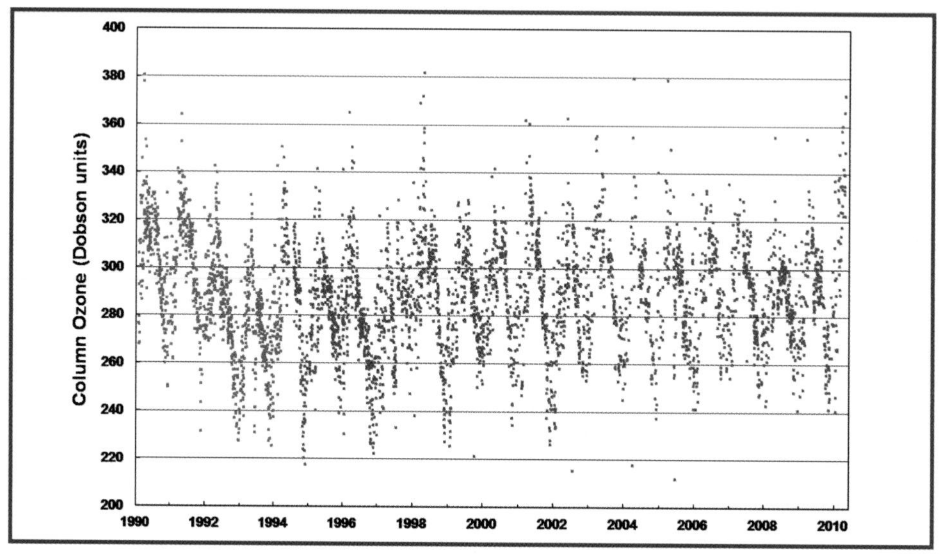

FIGURE P-5. The ozone layer over South Texas, measured by the author. Red points from 1990 to 1994 were measured by TOPS-1. Blue points from 1994 to 1997 were measured by Microtops and Supertops. Points from 1997 to 2010 were measured by Microtops II, manufactured by Solar Light.

1 How to Study Tree Rings

In this chapter you'll find projects that hopefully will encourage you to begin doing science, whether you're a student looking for a good science fair project or an adult wanting to start personal science study.

Tree Rings

In temperate and arctic regions, most trees are dormant during winter. When spring arrives, a sudden burst of growth expands trunks and branches with new wood, known as early wood (or spring wood), formed from large cells.

As the growing season peaks, the growth slows, and the late wood (or summer wood) that forms has cells with thicker walls. This late wood may appear much darker than the early wood when it contains more tannin (Figure 1-1).

FIGURE 1-1. Cross-sections from two varieties of bald cypress felled by a Texas flood.

Each year, this process forms a new growth ring, just beneath the bark of the trunk and branches of a tree.

Not all trees produce annual growth rings. Trees in the tropics that grow year round may have very suppressed annual rings or none at all. I learned this firsthand while sampling trees in Brazil during a field trip sponsored by NASA to measure the impact of severe biomass smoke on the atmosphere and plants.

Where I live in Texas, most trees are dormant during winter. But only some of these trees produce sharply defined rings. These include the red oak, hackberry, and all pine and bald cypress trees (Figure 1-2). The live oak keeps its leaves during the winter, and its rings can be difficult to count.

FIGURE 1-2. Narrow rings in this bald cypress accompanied a 1950s drought; ring chart by ImageJ software.

The Science of Tree Rings

Astronomer A. E. Douglass established tree ring science when he postulated that tree growth was influenced by climate changes caused by the solar cycle. He developed and taught classes at the University of Arizona on dendrochronology, the science of dating trees by studying their rings. In 1937, he established the university's Laboratory of Tree- Ring Research.

Douglass showed that archaeologists could use growth rings to date the timbers used to build ancient structures. Annual growth rings in trees also provide valuable information about past precipitation, climate, major volcano eruptions, and forest fires (Figure 1-3). They permit long-ago floods and landslides to be dated.

FIGURE 1-3. Century-old Norway spruce felled by chainsaw.

How to Obtain and Prepare Tree Ring "Cookies"

You can learn to do tree ring studies by using slices, or "cookies," cut from trunks and branches. Christmas trees are an excellent source, as are building and road construction sites. Professional tree trimmers and landscape crews may be willing to provide you with samples. Firewood may also provide good sample material. Another way to find samples is to keep an eye out for piles of recently cut and discarded branches.

You can even use cross-sections sawn from lumber, although these will not form round cookies. If you use this method, try to find dated lumber that includes the outer edge of the original trunk so you can determine the age of the rings.

If you have trees or access to trees where you live, you can cut branches or collect cores from trunks. If you're not the landowner, be sure to get permission first. This is especially important if you want to obtain samples from trees on private land or land owned or managed by cities, states, or the federal government.

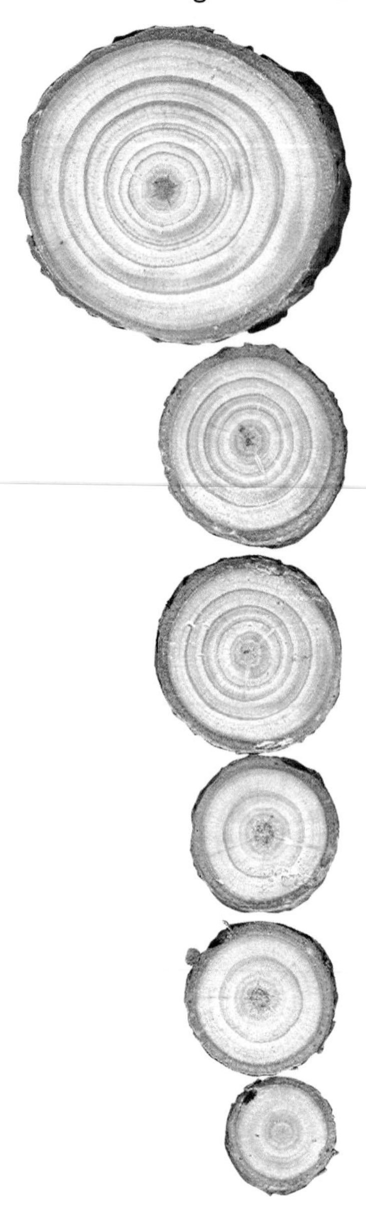

FIGURE 1-4. This sequence of "cookies" cut from a branch at the top of a fallen pine tree allows its growth to be carefully analyzed.

When possible, use a sharp, fine-toothed wood saw to slice cookies from branches and trunks.

Living wood should be allowed to dry for a day or two before smoothing it with sandpaper. I usually begin with 100-grit sandpaper followed by 220 grit. The final polish is made with 400 or 600 grit.

Samples cut with a chainsaw can be used, but they'll require much more surface preparation. If possible, smaller samples cut with a chainsaw should be recut with a handsaw. You can use a power sander to smooth the rough faces of these samples.

For small samples, I prefer to use a handheld plane such as the Stanley 21-399 5-Inch Surform Pocket Plane. This tool will quickly remove burrs and other saw marks.

Large trunk cross-sections require considerable time to prepare (Figure 1-4). A local cabinet shop once smoothed some large bald cypress sections that I had used a chainsaw to remove from trees knocked down by a major flood. But the cabinet shop couldn't handle the largest section, which was more than three feet across.

How to Examine and Photograph Tree Ring Samples

Your sample is ready when it's been sanded smooth to the touch and has few remaining saw marks. If the polished side of the sample looks good, flip it over and use a ballpoint pen or fine-point Sharpie marker to write the species and the date and place where it was collected.

It's best to examine the rings with a magnifying lens or a 10x loupe. Note that individual rings may be dark on the side nearest the bark and light on the side nearest the center. Be careful to count these two differences in shading as one ring and not two (Figure 1-5).

FIGURE 1-5. This Norway spruce log was used to build a cabin in Switzerland. Experienced dendrochronologists can compare these rings with those of logs with known dates to determine when this log was cut.

The first thing I do is count the rings by their year, beginning from the first one inside the bark. You may want to place a mark at each tenth year. Ideally, try to determine the date when the branch or trunk began growing. After you determine this date, print it on the backside of the sample.

Professional tree ring analysts use various stains to highlight rings that are faint and difficult to see. You can even use water. Just moisten a paper towel in water and lightly stroke it across the sample.

FIGURE 1-6. You may need permission to take tree "cookies" across international borders, so take photos.

You can make photographs or digital scans of your tree ring samples for more detailed analysis and for display online. I've used scanners and digital cameras with a close-up setting. Try moistening samples as described above to enhance visibility.

Because of international travel restrictions, if you collect samples outside your country it's best to leave them behind and bring home only their photos (Figure 1-6).

You can use various photo processing programs to further enhance the visibility of the rings. And you can use ImageJ and other image analysis tools to help count the rings and study their color differences. ImageJ requires no license, and the program is freely available at http://rsbweb.nih.gov/ij/index.html. (See Chapter 12, "How to Analyze Scientific Images" for more about ImageJ.)

Going Further

There are some excellent websites devoted to tree rings. By far the most comprehensive is Dr. Henri D. Grissino-Mayer's Ultimate Tree-Ring Web Pages at http://web.utk.edu/~grissino/. This site includes a superb collection of tree ring images, background information, tips, and links to other tree ring sites (Figure 1-7).

Tools known as increment borers are used to extract cores of wood from living trees without having to cut down the tree. Suppliers of these tools are listed on the Ultimate Tree-Ring Web Pages. In my experience, an increment borer can never replace a full cross-section, but they're invaluable when cross-sections of trunks are simply unavailable. These tools are much easier to use in conifers than in hardwood trees, as I found out while working up a sweat coring hardwoods in Brazil's Amazon basin.

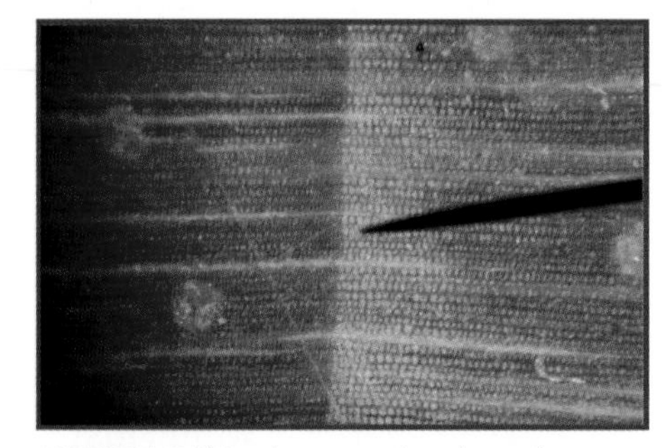

FIGURE 1-7. This microscope view shows the boundary between two growth rings in a pine branch cut by landscapers at Mauna Kea State Park in Hawaii. Growth is from left to right.

2

Snow Science

Depending on the circumstances, a landscape coated with snow can be a winter wonderland, a major nuisance, or a disaster in the making. There's another way to view the white stuff as well because snow can provide an important resource for doing science.

Both scientists and photographers have long studied and photographed the astonishing beauty of individual snowflakes. Here we'll concentrate on snow that has reached the ground. We'll explore some of its characteristics and effects on the environment, aside from the moisture it provides.

Snow as a Heat Island Indicator

The temperature measured by many of the ever-diminishing number of climate monitoring stations around the world is biased in the warm direction by improper site selection or by changes at sites that were properly located when they were first installed.

In 2007, meteorologist Anthony Watts became concerned about this problem and began a project to survey all the climate monitoring stations in the United States Historical Climatological Network (USHCN). Watts' project has so far provided photographs and detailed descriptions of 1,003 of the 1,221 USHCN stations surveyed by Watts and his volunteer team of citizen scientists.

The most alarming finding from this study is that many stations are placed much too close to "heat islands" such as buildings, pavement,

FIGURE 2-1. Sapling conifers like this one near Cloudcroft, N.M., form heat islands that melt surrounding snow, especially when sunlight warms their needles.

sidewalks, driveways, and even the hot exhaust from air conditioners. Only 10 percent of the stations surveyed meet the National Oceanic and Atmospheric Administration's (NOAA) two highest rankings for climate stations. Full details of Watts' project are at http://surfacestations.org/.

Watts and others have used expensive infrared viewers to see how weather stations are influenced by nearby heat islands. Snow can also indicate a warming bias, which can be easily recorded by an ordinary camera.

There is likely a problem if snow melts more rapidly in the vicinity of a temperature station than in a nearby open area. The temperature-sensing apparatus itself and its mounting hardware can provide a slight warm bias. But the most significant warming is often caused by nearby roads, parking lots, and buildings.

Snow also provides an ideal tool for photographing and studying natural heat islands. For example, rocks emerging through snow will quickly warm when exposed to sunlight and melt nearby snow. Stumps and living vegetation will also become warm and cause melting. This provides interesting clues about the survival of insects and microorganisms during winter.

FIGURE 2-2. This rock near Sunspot, N.M., formed a natural heat island when warmed by sunlight. Note the leaves peeking out from the ice.

You can do a variety of experiments that illustrate how heat islands affect snow. The simplest is to place various objects on snow in an open area with plenty of sunlight. Sheets of black and white construction paper will work if there is no wind. Take photos of the objects and the surrounding snow before and after the sun has done its work.

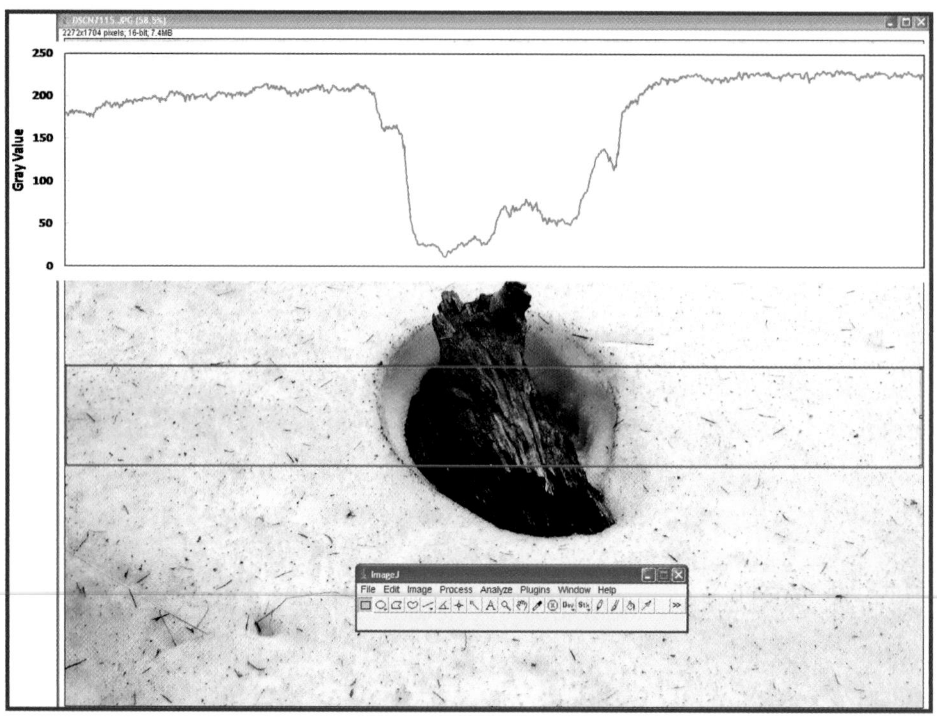

FIGURE 2-3. The heat island effect formed by this burnt log can be outlined using ImageJ image analysis software (see Chapter 12, "How to Analyze Scientific Images"). The plot indicates the snow with high values and the burnt wood with low values.

Snow as a Particle Collector

During the spring of 2004, fires in Southeast Asia sent smoke plumes across the Pacific Ocean to the United States. When the Navy Research Lab's NAAPS aerosol forecast (www.nrlmry.navy.mil/aerosol/) showed that smoke would arrive in New Mexico, I headed west from Texas in my old pickup in an effort to both measure the smoke and capture some of it for study. Along the way, a homemade air sampler mounted on the truck blew air collected by a funnel over the sticky side of a piece of adhesive tape.

The air sampler proved its worth by collecting mineral grains, fungal spores, pollen, and other matter while driving along West Texas and New Mexico highways. But it failed to collect any of the microscopic particles of coal-black soot that form smoke. The Asian smoke remained high overhead and none of it fell to the surface while I was sampling. The smoke covered much of New Mexico, but when I arrived in Las Cruces on March 25 it was much too high to be captured.

Finding particles of smoke that might have fallen from the smoke clouds before I arrived in New Mexico would require a very different kind of collector, so I headed for Cloudcroft in the nearby Sacramento Mountains. The NAAPS forecast showed that smoke particles fell to the ground over that region on March 16–17.

Large patches of snow remained in the mountains, and they were coated with a surprising amount of dust. The snow under the dirty layer was much cleaner. The NAAPS model showed that a large dust storm had blown across the mountain on March 5.

In effect, the patches of snow served as giant air samplers that captured and stored whatever was falling from the sky. While smoke particles were the main objective, the dust provided a bonus. It was time to place a drop of melted snow on a glass slide and inspect it through a microscope.

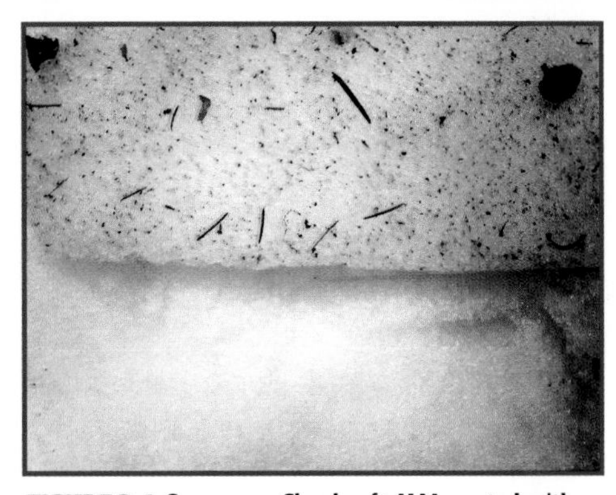

FIGURE 2-4. Snow near Cloudcroft, N.M., coated with dust from White Sands and smoke from Southeast Asia.

A single drop of snowmelt from the dirty layer contained hundreds of fungal spores and many thousands of tiny grains of gypsum. Also present were plant matter, transparent orange crystals, and what appeared to be a few slivers of volcanic glass.

FIGURE 2-5. Samples of dirty (left) and clean snow (right) collected near Cloudcroft, N.M.

Dozens of samples were inspected, and many included dark black particles of soot scattered among the tiny grains of gypsum dust and sand. The soot was presumably from the Southeast Asian smoke that fell across the region the week before.

The obvious question about the fungal spores was, did they arrive with the dust or the smoke?

Or maybe they blew from nearby evergreen trees?

The smoke hypothesis can't be ruled out. When my daughter Sarah was in high school, she discovered many spores and bacteria in biomass smoke that arrived in Texas from the Yucatan Peninsula. She established this beyond a reasonable doubt by capturing spores arriving with smoke from the Yucatan by means of a homemade air sampler she flew from a kite at the edge of the Gulf of Mexico.

FIGURE 2-6. A drop of dirty evaporated snowmelt on a microscope slide left behind this deposit of dust, smoke particles, fungal spores, and plant matter.

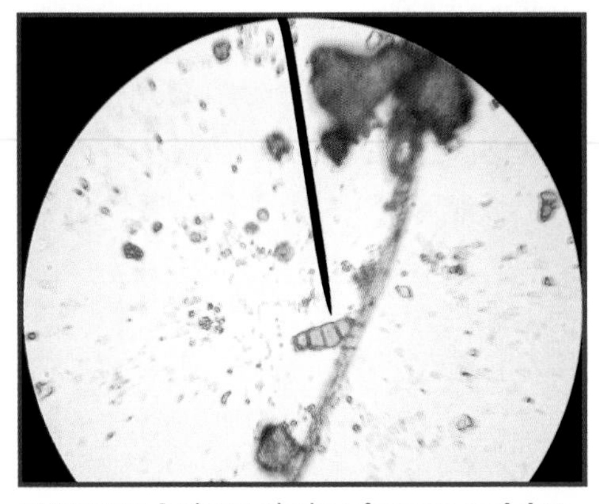

FIGURE 2-7. A microscopic view of gypsum sand, dust, plant matter, and spores from snow near Cloudcroft, N.M. The pointer indicates a Curvularia spore.

This discovery became a "fast track" paper in *Atmospheric Environment,* a leading scientific journal (see http://earthobservatory.nasa.gov/Features/SmokeSecret/smoke_secret2.php). While spores can probably be blown across the Pacific, it's much more likely that those in my snow samples came from much closer sources, such as nearby trees. In any event, it was quite a surprise to find so many fungal spores in snow atop a mountain. The spores included the genuses Alternaria, Nigrospora, Curvularia, Cladosporium, Penicillium (or possibly Aspergillus), and what resembles the ascospores Splanchnonema and Leptosphaeria.

I was unable to identify the most common spore in the snowmelt, as it's not shown in E. Grant Smith's *Sampling and Identifying Allergenic Pollens and Molds* (Blewstone Press, 2000) or Bryce Kendrick's *The Fifth Kingdom* book and CD (Mycologue Publications, 2000).

Going Further

If you reside in snow country, what's in your snow? Does apparently pristine snow contain fungal spores and protozoa? Is it contaminated with soot particles? If these and other contaminants are present, can you use satellite imagery to backtrack the storm that dropped the snow, thereby possibly finding where the contaminants originated?

How do plants buried under snow exploit the heat islands formed by rocks and stumps?

FIGURE 2-8. A fresh blanket of snow has arrested the melting of snow sprayed with sand spread by a road crew atop New Mexico's Sandia Mountain.

What are the simplest and most efficient ways to melt snow outdoors? Ashes? Sand? Sunlight reflectors? Reusable black plastic sheets?

Finding answers to these and other snow questions can lead to a variety of intriguing science projects for those who reside in or who visit snow country.

3 Tracking Heat Islands

Accurately monitoring the temperature outdoors isn't easy. That's because air temperatures are influenced by virtually anything heated by sunlight, and by engines and other equipment that generate heat.

Bubbles of warmth are known as heat islands. They make major contributions to the temperature in and around towns and cities, which are usually warmer than the countryside. The National Weather Service has developed guidelines for placing weather instruments to avoid heat islands (see www.srh.noaa.gov/epz/?n=cwopepz). But with the passage of time, fewer weather stations meet the guidelines, often because of the arrival of new buildings, sidewalks, and paved roads.

Electronic Thermometers

Stand-alone temperature loggers are available from Onset Computer, Lascar, Extech, and others. Data are downloaded through USB or optical ports; software converts the data into graphs.

You can modify an Onset Pendant logger for temperature transects by boring an entry hole through its cap and an exit hole in its base, to allow air to flow over the thermistor, a tiny resistor whose resistance varies rapidly with temperature.

While temperature loggers work well, they require you to note times and locations during a transect so you can analyze the data. No notes are needed with a Vernier LabQuest 2 (Figure 3-1). After making a transect, you can send up to 1,000 measurements to Vernier, which will return a Google map with a color-coded line that indicates the temperature along the route. While the LabQuest 2 is expensive ($329), mine has become essential.

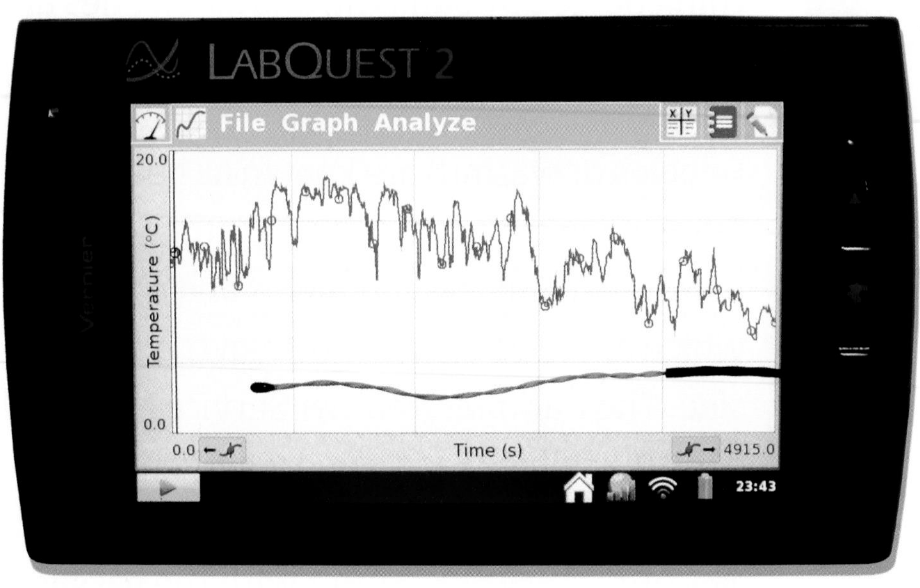

FIGURE 3-1. This 11-bit, fully programmable logger has three analog inputs for external sensors, and a built-in GPS that records the location of every measurement. Best of all, it displays a real-time graph of temperature as it is measured.

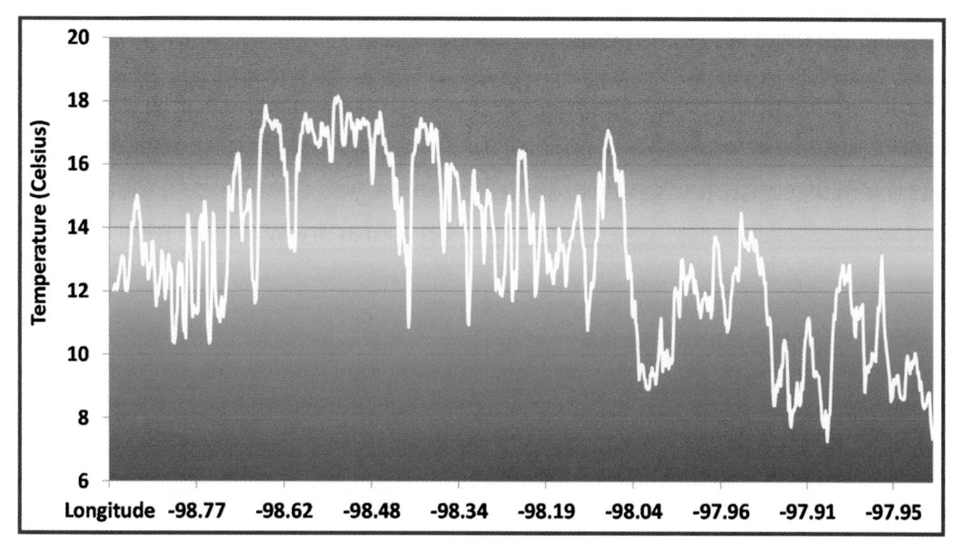

FIGURE 3-2. Excel plot of 942 nocturnal temperature measurements across San Antonio, logged with my LabQuest 2 and a thermistor installed on my pickup.

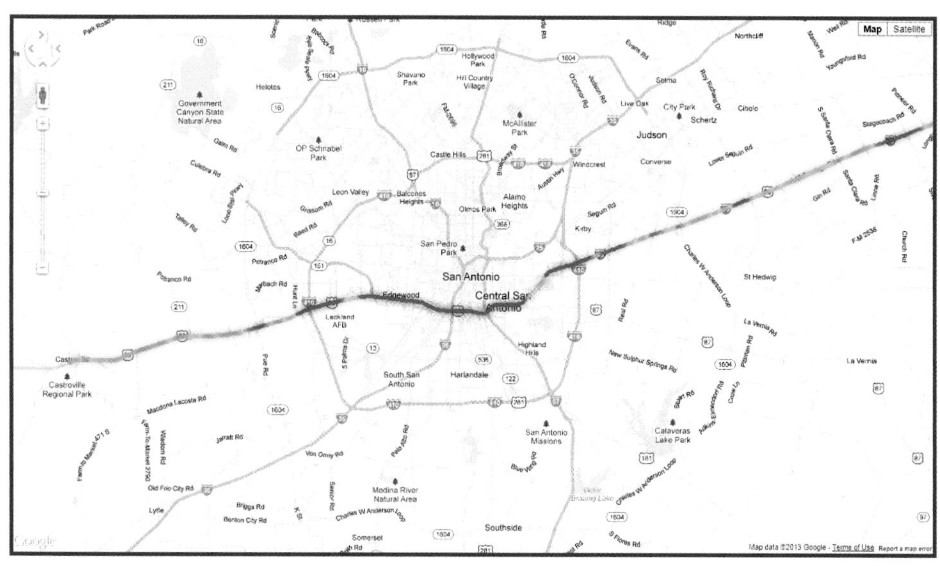

FIGURE 3-3. The same data as Figure 3-2, on the map generated by Vernier.

Suitable Temperature Sensors

Dedicated temperature loggers have a built-in sensor. The LabQuest 2 works with many external sensors, including a Vernier temperature sensor ($23). If you already have a voltage logger, you can use it to measure temperature with the help of a thermistor.

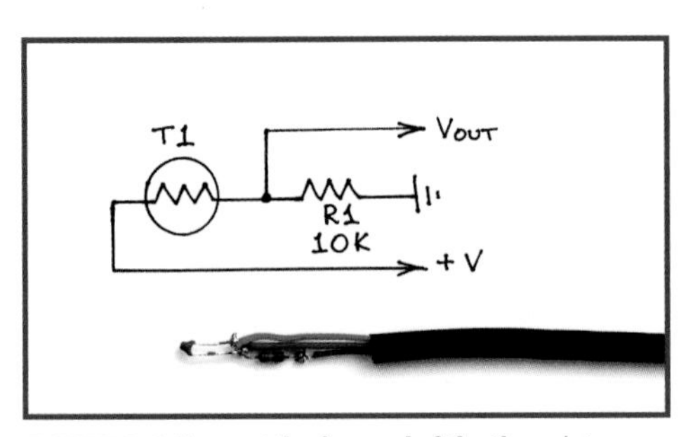

FIGURE 3-4. Connect the free end of the thermistor to the logger's positive voltage and the free end of the resistor to ground. The voltage between the thermistor-resistor junction and ground will be directly proportional to the temperature.

You can make a DIY temperature sensor by connecting a 10kΩ thermistor (Jameco #207037 or similar) in series with a 10kΩ resistor to form a voltage divider (Figure 3-4). You can swap the resistor for a 20kΩ–50kΩ potentiometer to adjust the sensitivity of the probe.

Mounting Temperature Sensors

When mounting temperature loggers on a car, remember that the temperature sensor must be shielded from sunlight. If the car will need to make stops at traffic signs or lights, the sensor should be mounted away from the vehicle to avoid heat buildup from the engine or metal surfaces. Here are some methods I've used:

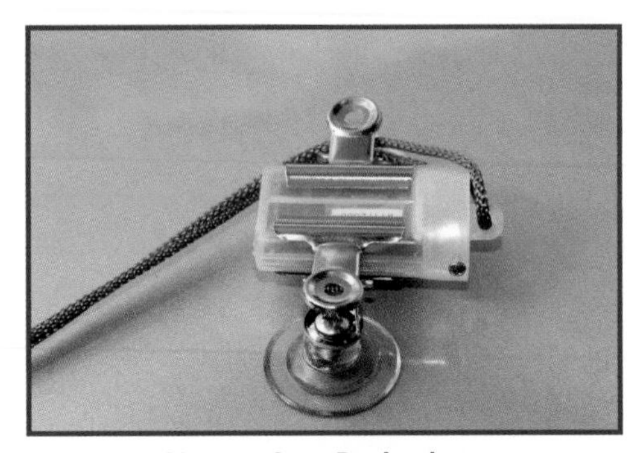

FIGURE 3-5. Mount an Onset Pendant logger on a car roof with a pair of Adams suction cup clamps and a safety line.

- Install an external thermistor in a drinking straw or paper tube taped to the roof (Figure 3-6).
- Use a car-window flag mount for an external sensor (Figure 3-7).

FIGURE 3-6. Insert the sensor into the tube's forward opening, tape its cable to the door frame, and run the cable through the window.

FIGURE 3-7. Remove the cap and the flag. Extend the sensor just beyond the upper end of the flagpole and secure it by wrapping a few inches of insulated, solid wire around the sensor leads and the groove at the top of the flag mount. Use tape or binder clips to secure the cable to the flagpole.

Make sure your temperature sensor is securely mounted to your vehicle and doesn't pose a hazard to other drivers or become a distraction to your driver. Come to a complete stop in a safe area before making any changes to the experiment.

- Use 2 feet of ½" PVC pipe with a tee fitting to make a sturdy sensor mount that's shielded from sunlight (Figure 3-8).

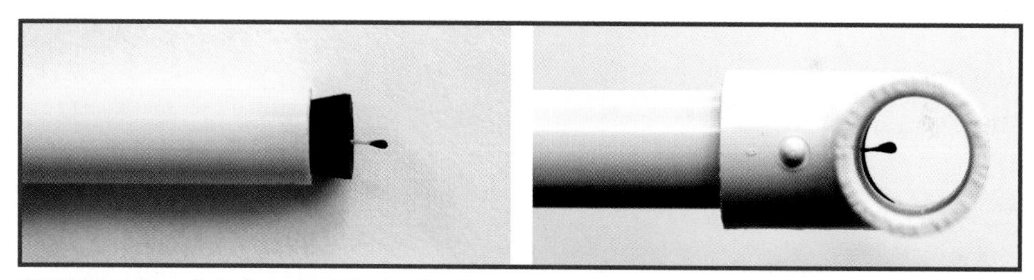

FIGURE 3-8. Bore a hole through a ¾" × 9/16" rubber stopper and push the sensor through so that it protrudes 5/16" from the large end. Insert the sensor cable into one end of the pipe and press the stopper in. Bend the free end of the cable back toward the sensor and tape it securely to the pipe. Finally, place a ½" tee over the end of the pipe so the sensor is visible through both ends.

Vibration Sensors

4

Many methods have been improvised, invented, and hacked to sense sudden movements, acceleration, and vibration. Today tiny, solid-state accelerometers are available to do these jobs, and they're embedded in many smartphones, tablets, and electronic game controllers. These miniature accelerometers are available for reasonable prices, or you can use the smartphone itself for vibration detection.

DIY vs. Solid-State Vibration Sensors

There's no need for specialized accelerometer circuits if all your project needs is a simple vibration detector. Many DIY methods are available—and some are more sensitive than the solid-state accelerometers in smartphones.

This can be demonstrated with an Einstein Tablet educational computer (http://einsteinworld.com/home/). The Einstein can store data from up to 16 sensors, including its own internal 2-axis accelerometer and photodiode. These two sensors allow us to directly compare a solid-state accelerometer with a DIY vibration sensor made of a light source and the Einstein's photodiode.

First, transform the Einstein into a pendulum by suspending it from a tabletop with 2" shipping tape. The photodiode (the dark square sensor in Figure 4-1) should face down and the screen should face away from the table. Place an LED flashlight on the floor and point it up at the photodiode.

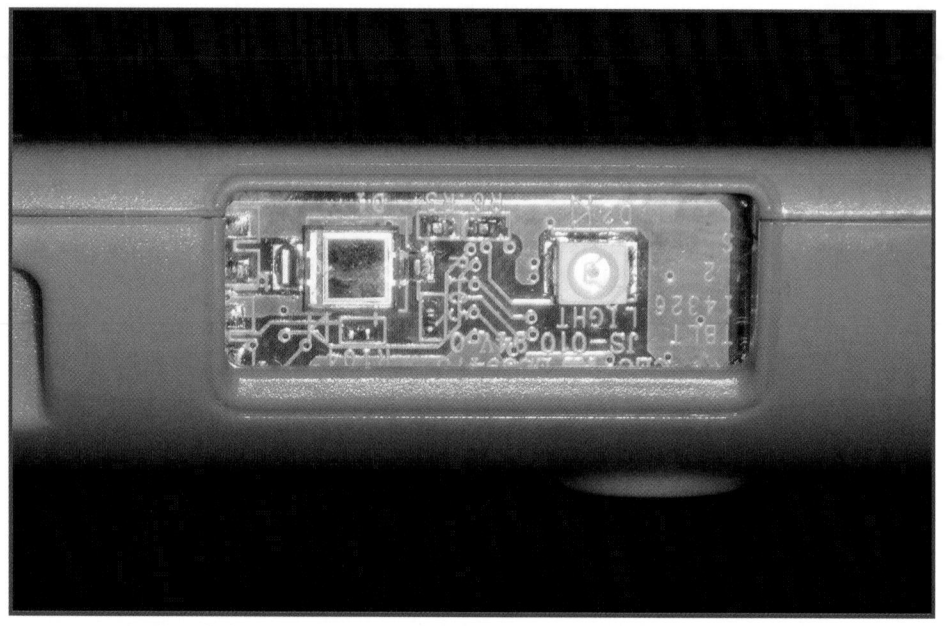

FIGURE 4-1. The photodiode on the edge of an Einstein Tablet.

Select the accelerometer and photodiode options by checking them in the Einstein's launch window. Set the sample rate at 10 per second, the sampling duration at 50 seconds or more and the sensitivity of the photodiode at 0–600 lux. Press the Start arrow and then pull the Einstein 1" or so away from the vertical. Then release the tablet so that it swings back and forth. Experiment with the placement of the flashlight for best results.

Figure 4-2 is a chart comparing the movement response of the z-axis of the Einstein's accelerometer and its photodiode. The photodiode is plainly more sensitive to movement, and even shows a periodic wobble in the swing of the tablet.

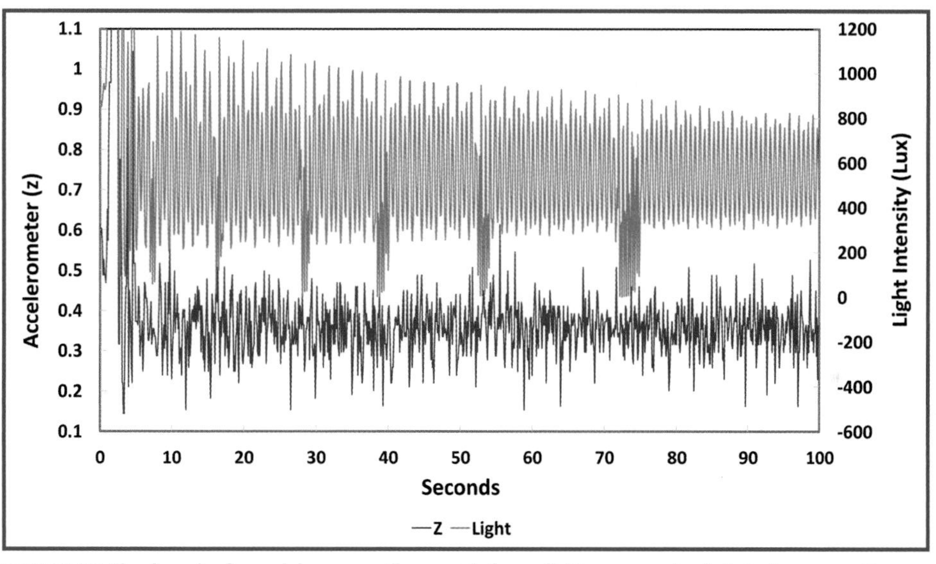

FIGURE 4-2. Clearly, a simple pendulum or cantilever made from a light source and a photodiode can provide a very sensitive movement and vibration detector.

Make a Piezoelectric Vibration Sensor

Here's another DIY method that's simple and doesn't need to be shielded from external light. A piezoelectric crystal or ceramic generates a voltage when it is bent or struck. A very simple vibration sensor can be made from the type of piezo ceramic disc that emits tones and sounds in watches, phones, greeting cards, and alarms.

Figure 4-3 shows a circuit that flashes an LED and emits tone bursts when piezo disc PZ1 (mine was salvaged from a greeting card) is touched or vibrated.

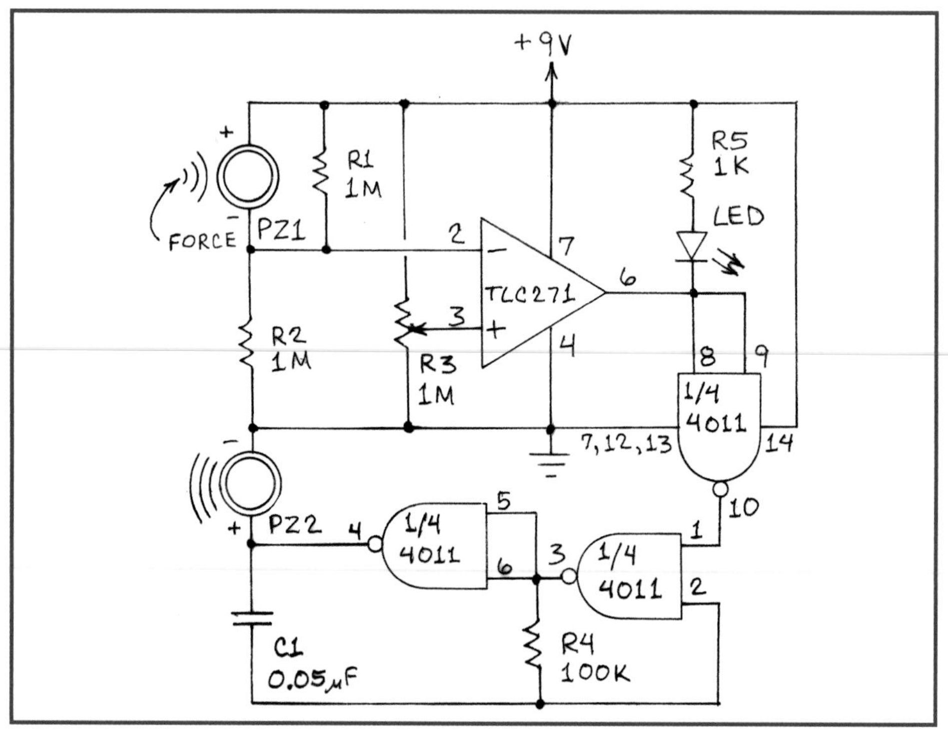

FIGURE 4-3. The piezo disc is connected to the inverting input of a TLC271 or similar operational amplifier connected as a voltage comparator. A voltage divider formed by potentiometer R3 is connected to the noninverting input of the op-amp.

In operation, R3 is adjusted until the output of the op-amp switches from low to high. This switches off the tone generator formed by a 4011 quad NAND gate. A very slight mechanical shock will cause PZ1 to generate a voltage that will switch the comparator output from high to low. This will flash the LED and cause the 4011 to generate a tone burst.

This circuit can be built on a solderless breadboard (Figure 4-4) that allows for easy modifications. For example, the tone frequency can be decreased by increasing the value of C1. The sound volume can be increased by replacing PZ2 with the input side of a standard audio output transformer having an input impedance of 1,000Ω and an output impedance of 8Ω (RadioShack 273-1380 or equivalent). Connect a small 8Ω speaker to the transformer's output. For even more volume, connect pin 4 of the 4011 and ground to the input of an external amplifier.

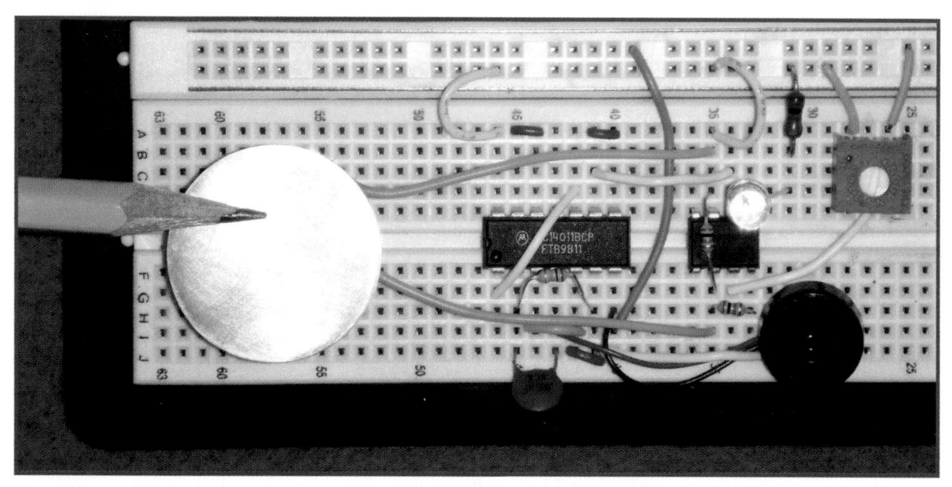

FIGURE 4-4. A Piezoelectric Vibration Sensor built onto a solderless breadboard.

The sensitivity of the circuit can best be demonstrated by connecting PZ1 to the circuit with a pair of 4", 24-gauge jumper wires as shown in Figure 4-4. Solder one end of each wire to the connection points on the backside of the piezo disc. Connect the other two ends of the wires to the circuit so that the piezo disc is cantilevered, suspended about 1" over the breadboard.

When PZ1 is still, adjust R3 until the LED glows and the tone generator is on. Then back off on R3 until both the LED and the tone are off. Now when you barely touch the piezo disc, the LED and tone generator should respond (Figure 4-5).

FIGURE 4-5. Tap the piezo disc so that it bounces up and down, and the LED and tone will respond accordingly.

PZ1 can be kept in the cantilever mode or cemented or clipped to a wall, curtain, step, car, etc. The circuit can be made much more sensitive by cementing to its upper surface a stiff metal rod with a small weight mounted on its free end. Footsteps can be detected when a piezo disc with an attached rod is mounted on or under a flat surface such as a wood floor or step.

Going Further

Sensitive pendulum and cantilever vibration sensors can be made from hardware found in almost any workshop. Suspend a weight from the end of a vertical metal rod or mount it on the end of a horizontal metal cantilever. Mount a machine screw and nut near the movable weight so the screw can be adjusted to nearly touch the weight or its rod. You now have a movement and vibration sensitive on-off switch.

Make an Experimental Optical Fiber Seismometer

5

Volcano eruptions, slippage of faults, explosions, landslides, drilling, and even traffic can create vibrations in the earth's crust known as seismic events or earthquakes. Seismometers are devices that detect seismic events, usually by mechanically or electronically detecting the movement of a suspended mass. Miniature accelerometers inside smart phones and video games employ tiny movable masses to detect movements, including seismic events.

Optical Fiber Seismometer

My son Eric Mims developed a novel optical fiber seismometer when he was in high school. His system was an optical fiber pendulum suspended from a steel frame bolted to the concrete slab under the carpet in his bedroom. A weight was attached to the free end of the fiber, which hung directly over an LED mounted under a pinhole. Seismic events caused the end of the fiber to oscillate back and forth across the pinhole. A photodetector at the opposite end of the fiber detected changes in light intensity, which were amplified and sent to a printer that Eric programmed as a chart recorder. This simple device detected earthquakes and two underground nuclear tests in Nevada, which earned Eric a record number of awards at the Alamo Regional Science and Engineering Fair in San Antonio.

Materials

- IC1—TLC271 operational amplifier
- LED—super bright red LED
- PD—BPW34 Photodiode (Jameco Part No. 1621132 or similar)
- R1—3 to 5 megohm resistor (higher values give more sensitivity)
- R2—10,000 ohm resistor
- Solderless breadboard
- Plastic optical fiber (Jameco 2.2-mm fiber or similar)
- Lead or steel fishing sinker with central hole
- Data logger (optional; see text)
- Hot melt glue gun

Build an Experimental Optical Fiber Seismometer

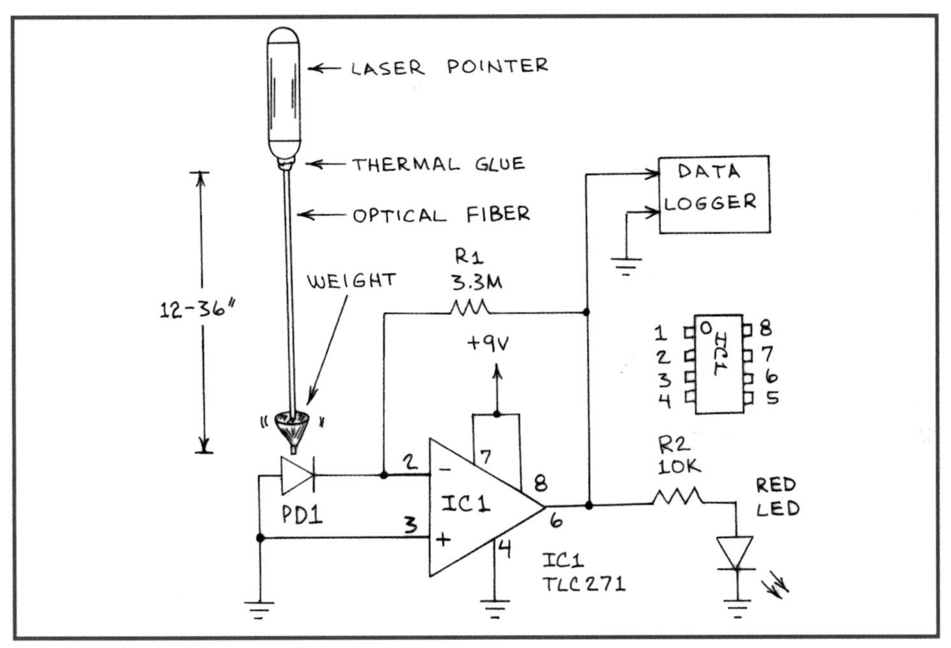

FIGURE 5-1. The circuit diagram for the detection section of Eric's fiber optic seismometer.

Figure 5-1 is an outline view and circuit diagram of a revised version of Eric's fiber optic seismometer made from a meter or so of fiber, a fishing weight, a laser pointer, and a photodiode connected to an amplifier. One end of the fiber is inserted into the open end of a laser pointer and secured in place with a hot melt glue gun. The opposite end of the fiber is inserted through the central hole in a bullet-style fishing weight (Figure 5-2).

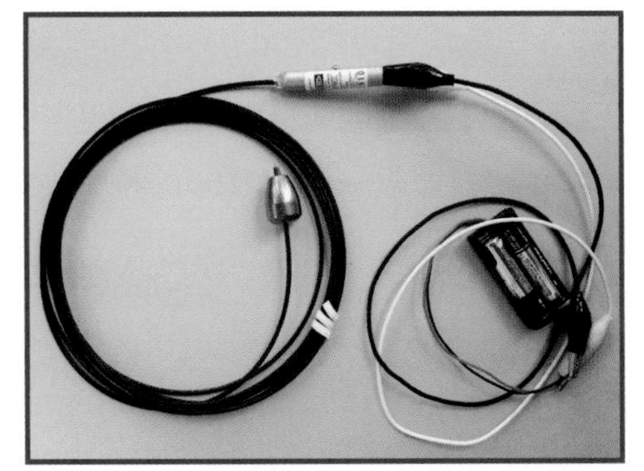

FIGURE 5-2. The emitting section of the seismometer, without the rigging: a power pack, a laser, and a coil of optical fiber strung through a fishing weight.

The detector circuit is built on a solderless board as shown in Figure 5-3. The laser light emerging from the weighted end of the fiber is detected by a photodiode or miniature silicon solar cell. A red LED glows to indicate when the solar cell is receiving light, so it's necessary to test the system in a darkened room.

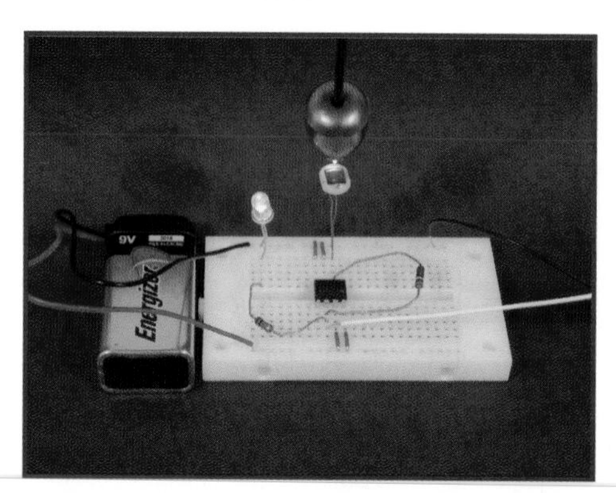

FIGURE 5-3. The photocurrent from the cell is amplified by a TLC271 operational amplifier, the gain of which is determined by the resistance of R1. The higher the resistance, the more sensitive the circuit is to light.

For initial tests, place the laser end of the fiber on a table and suspend the weighted end over the table's edge. Place the detector circuit on the floor, and adjust the fiber and the circuit until the end of the fiber is suspended just over the solar cell. When the laser is switched on, the detector circuit's red LED should glow. Gently pushing or even blowing on the weight will cause the fiber to oscillate back and forth like a pendulum. Each time the end of the fiber sweeps over the solar cell, the LED will glow.

To detect very subtle movements, move the circuit board until the end of the fiber is over the edge of the detector. Or make a pinhole in a square of black tape and place the tape over the photodiode.

If you plan to detect earthquakes, install the system inside a dark enclosure or a closet on the concrete slab or basement of a house. The LED can be connected to the circuit with wires long enough to allow it to be mounted outside the enclosure. The laser pointer will quickly exhaust its miniature batteries, so remove them and connect a 2-volt battery holder with a pair of AA or AAA cells to the pointer using alligator clip leads. If the laser employs a push button switch, close it with tape or a clothespin.

You can connect a data logger to the seismometer to provide a record of what it detects. The data logger used for Figure 5-4 is a 3-channel, 15-bit Hobo UX120-006M Analog Data Logger made by Onset Computer Corporation.

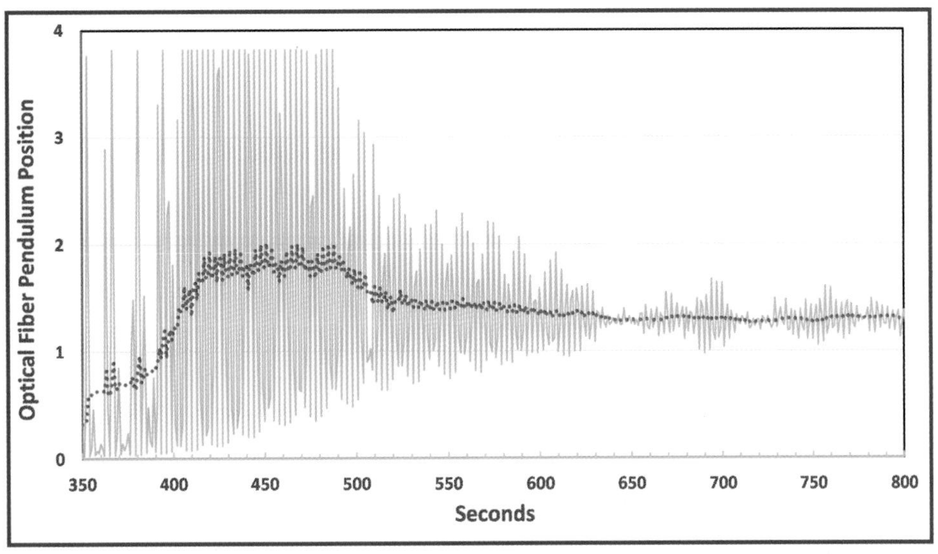

FIGURE 5-4. A typical response of the seismometer when its weight is pushed slightly and allowed to swing until it is again stable.

Modifying the Seismometer

You can make many modifications to the basic optical fiber seismometer. For example, you can connect a piezo buzzer or tone generator to the output of IC1 to provide an audible signal when movement is detected. You can detect the in-line direction of the seismic event by replacing the photodiode with a quadrant photodetector. Center the free end of the fiber at the junction of all four quadrants of the sensor. Connect each quadrant to its own opamp and their outputs to a 3-input data logger like the Hobo. You can make your own quadrant detector from four miniature solar cells cemented to a base.

Dampening the Seismometer

While this seismometer is ultra-sensitive, it takes time to settle back to its neutral position after a seismic event. Dampening the pendulum reduces its sensitivity, while also reducing the time required for the pendulum to settle back to neutral after a vibration event. Therefore, this improves the time resolution of the device.

The simplest way to dampen an optical fiber seismometer is to simply shorten the fiber. Another way is to place a clear container with a flat bottom over the light sensor. Fill this container with clear vegetable oil and dip the weighted end of the fiber into the oil. The amount of damping is determined by the depth of the oil.

A Video Optical Fiber Seismometer

The simplest way to determine the magnitude and direction of movement of an optical fiber pendulum during a seismic event is to place a video camera under the end of a fiber from which light emerges. Figure 5-5 is a 15-second time exposure of an oscillating optical fiber pendulum. A video was made of the moving end of the fiber with a Panasonic Lumix camera in video mode. The video was played on a Surface Pro 3 tablet, and a 14-second time exposure of the video was made with a Sony a6000 (ISO 100 at f7). When Eric saw this image, he suggested using a webcam that records only when movement is detected. That's definitely on my list of things to do.

FIGURE 5-5. A 14-second time exposure of an oscillating optical fiber pendulum.

Going Further

You can learn much more about seismometry from online articles, including http://eqseis.geosc.psu.edu/~cammon/HTML/Classes/IntroQuakes/Notes/seismometers.html.

Various kinds of DIY seismometers are also available online, including www.mikesenese.com/DOIT/2010/04/photos-and-video-from-mexicalis-7-2-earthquake-and-how-to-build-your-own-seismograph/.

Ultra-Simple Sunshine Recorders

6

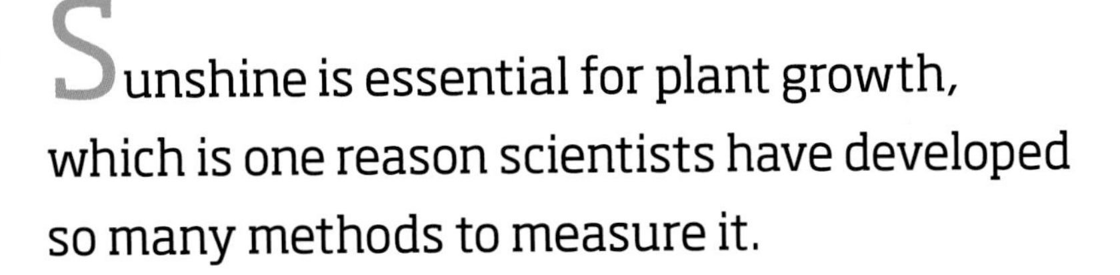

Sunshine is essential for plant growth, which is one reason scientists have developed so many methods to measure it.

In a future chapter I'll show how to measure sunshine electronically. But first, let's enter the *MAKE* time machine and zip back to 1838, when instrument maker T.B. Jordan made the first known automatic sunshine recorder.

Jordan wrapped a strip of silver chloride photographic paper around a clock-driven cylinder, and mounted it behind the mercury column of a barometer. The photo paper was exposed to sunlight as it rotated past an aperture, and this provided a record of when clouds blocked the sun. The width of the exposed portion of the paper was controlled

by the height of the mercury. Thus the instrument was both an early barometric pressure recorder and the first sunshine recorder.

In 1840, Jordan designed a second sunshine recorder whose photographic cylinder was exposed to sunlight passing through a pinhole in a second cylinder that rotated about the photo cylinder once in 24 hours.

You can make simple versions of Jordan's sunshine recorders using cyanotype paper. It's easy to use, and the image is preserved by simply immersing the paper in water.

The Cyanotype (Blueprint) Process

When he wasn't flying aircraft for the U.S. Air Force, my father was a civil engineer and an architect. Detailed architectural drawings, imaged on cyanotype paper and known as blueprints, were displayed on his office walls.

Today, blueprint means a detailed plan or a physical model of something, but originally it meant a cyanotype print of an architectural or engineering illustration. Blueprints were created by first making an original drawing on a sheet of translucent paper. This drawing was placed over a sheet of blueprint paper and exposed to ultraviolet light. The blueprint paper was then developed in water rendered slightly acidic. The paper exposed to the UV light became a rich blue, while the paper shadowed by the drawings became white.

The blueprint process was discovered by Sir John Herschel in 1842 and has remained essentially unchanged ever since. Herschel described his discovery in his paper "On the Action of the Rays of the Solar Spectrum on Vegetable Colors, and on Some New Photographic Processes," published in *Philosophical Transactions of the Royal Society* (June 1842). You can make your own blueprint paper or buy it from hobby and craft stores. I use SunArt paper by TEDCO (www.tedcotoys.com/). It's available in packages of 15 sheets measuring 4"×6", 5"×7", or 8"×10".

Ultra-Simple Pinhole Camera Sunshine Recorder

The simplest sunshine recorder is a light-tight enclosure in which blueprint paper is placed opposite a pinhole. This fits the definition of a true camera.

Before sunrise, the recorder is mounted in a fixed position so that the pinhole aims at a point about halfway between solar noon (where the sun will be at its highest point in the sky) and the southern horizon (northern horizon if you're in the Southern Hemisphere). After sunset, the recorder is opened and the blueprint paper is washed in water mixed with a few

FIGURE 6-1. Pinhole sunshine recorder made from a plastic 35mm film canister. A thin strip of masking tape holds the blueprint paper in place.

drops of lemon juice to enhance the color. The sun's track across the sky is preserved on the paper as a deep blue arc. Any interruptions in the arc indicate when clouds passed in front of the sun.

You can easily transform a plastic 35mm film canister into a sunshine recorder. Bore a ¼" hole in the side, then place a small square of aluminum foil tape over the hole, and use a needle to make a pinhole through the foil. Roll a 1¾"×22" piece of blueprint paper into a cylinder (sensitive surface facing inward) with a gap somewhat wider than the pinhole, then insert it into the canister so the pinhole is centered between the edges of the paper.

For best results, hold the paper in place with a thin strip of masking tape applied to the upper edge. Be sure to install the blueprint paper indoors to avoid exposing it to UV from sunlight. Then snap on the lid.

FIGURE 6-2. The film canister sunshine recorder can be mounted on a tripod by boring a ¼" hole in the lid and securing it to the tripod screw with a 3-40 nut.

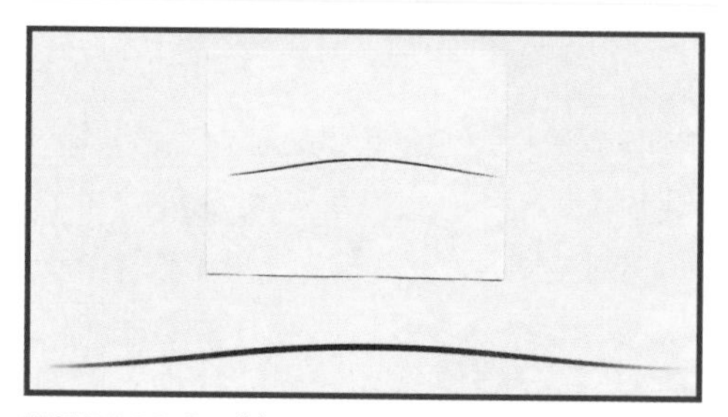

FIGURE 6-3. Sunshine traces made on the same cloud-free day by a 35mm film canister recorder with a 0.5mm pinhole and a tea can recorder with a 1mm pinhole.

Your pinhole sunshine recorder is now ready for use. Simply tape it to a fixed support so that the pinhole is pointed as described previously. For even better results, mount it on a camera tripod. This is easy; bore a ¼" hole through the canister's lid, then attach the lid to the threaded bolt of a tripod with a 3-40 nut. Then load the canister with blueprint paper and press it down into the lid.

If a pinhole recorder is left out in the rain, moisture can get inside. This shouldn't affect the previously formed sunshine trace, but it can cause the blueprint paper to wrinkle or the image to be prematurely preserved.

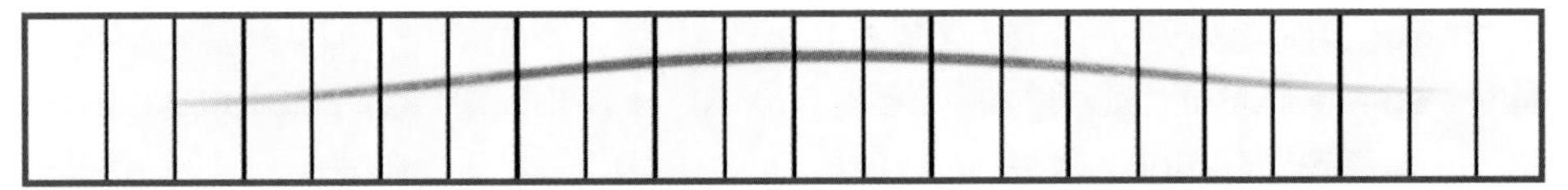

FIGURE 6-4. Digitized sunshine trace from tea can recorder in Figure 6-3 annotated with 30-minute grid lines. The center grid line was aligned with solar noon (12:36 p.m. on Jan. 2, 2011). Grid lines were equally spaced from noon to sunrise and noon to sunset. Daily sunshine times can be found at http://sunrisesunset.com/.

Clock-Driven Sunshine Recorder

Quartz-controlled analog clocks are available for as little as $10. You can easily make a sunshine recorder from such a clock. While this recorder uses a very small sunlight aperture, it's not a true imaging pinhole camera, since the aperture is in motion as it scans a speck of sunlight across a sheet of blueprint paper.

Start by removing the clock's housing; then remove the clock hands by carefully pulling them straight up from the face of the clock.

Cut a disk of black paper slightly smaller than the clock face, and make a ¼" hole at its center. Place the axis of the hour hand over the hole, and tape the hand to the disk. Make a 5" hole in the disk 2" from the center. This aperture will allow sunlight to strike the blueprint paper as the black disk rotates over it.

FIGURE 6-5. Quartz analog clock-driven recorder with two 1mm sunlight apertures. The black paper disk was taped to the clock's hour hand. Blueprint paper was taped to the clock face behind the black paper disk.

Cut a rectangular sheet of blueprint paper to cover the clock face's top half. Again, do this indoors so no sunlight or UV light strikes the paper. My clock face is 7¾" in diameter, so I cut a 3½"×5" piece of SunArt paper, with a V-shaped notch in the bottom to make room for the clock drive axis.

Place the blueprint paper over the clock face and secure it with masking tape. Finally, place the black disk over the clock face with its taped hour hand facing up, and its sunlight aperture aligned at a morning hour, and press the hour hand onto the clock axis. My 3½"×5" sheet allowed six hours of sunlight recording, so I set the aperture to the 9 a.m. setting.

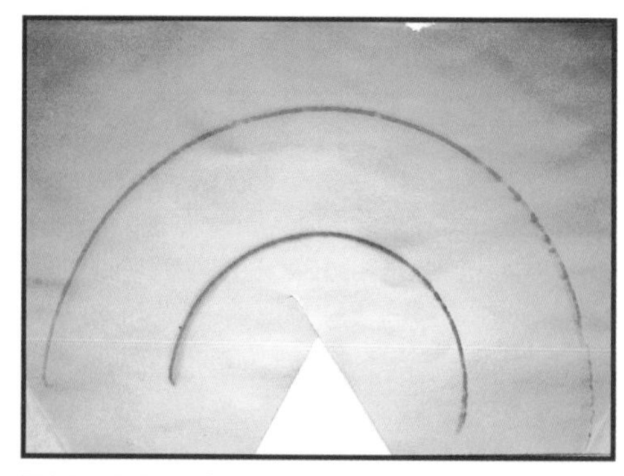

FIGURE 6-6. Dual sunshine traces made with the clock-driven sunlight recorder. Note the interruptions in the traces caused by clouds blocking the sun.

Place the clock outside so it faces the sky as described earlier. Retrieve it at sunset and remove the black disk, then remove the blueprint paper and preserve its image by immersing it in water with a few drops of lemon juice.

Going Further

To make your own cyanotype paper, see the instructions at www.howtodothings.com/hobbies/how-to-make-a-cyanotype. In Google Books, you can even find recipes and methods in Herschel's original paper.

Is the density of the blue color of a sunlight cyanotype trace proportional to the intensity of the sun's ultraviolet light? Finding out could be an excellent science project. Digitize a sunlight trace, then use photo software to determine the blue density at uniform intervals along the trace. Compare these data with the UV for your site measured by a Solarmeter (http://solarmeter.com/) or similar UV radiometer, or by the nearest radiometer in the U.S. Department of Agriculture's UVB monitoring program (http://uvb.nrel.colostate.edu/UVB/da_queryErythemal.jsf).

Finally, pinhole cameras can be used with ordinary photo paper to make spectacular sunshine traces of up to six months' duration. See Tarja Trygg's how-to instructions at www.solargraphy.com/ and Justin Quinnell's images at http://pinholephotography.org/.

The Infrared Thermometer: *An Essential Science Gadget*

7

If you do casual or even serious amateur science, you might already have some first-class science gadgets in your science tool kit. They might include a computer, a digital camera, and a thermometer. But do you have an infrared (IR) thermometer?

Your Science Tool Kit

Although I've spent 20 years measuring sunlight, haze, the ozone layer, and the water vapor layer using various homemade instruments, a couple of IR thermometers are among the most important gadgets in my science toolbox.

IR thermometers allow you to measure the temperature of objects without actually coming in contact with them. This is why they're called noncontact thermometers. They're ideal for basic chores like measuring the temperature of car engines, cooking surfaces, refrigerators, heaters, and streams and lakes. They also have many scientific applications.

IR Thermometers

Everything with a temperature above absolute zero emits IR radiation. IR thermometers use a thermal sensor such as a miniature thermopile to detect the IR emitted by objects. The signal from the sensor is corrected for the ambient temperature and sent to a digital readout. Some IR thermometers include an analog output for use with a data logger. Prices vary from $20 up to thousands. The Omega OS540 shown in this article cost $85; newer versions start at $95.

IR thermometers are usually employed to measure the temperature of surfaces, which may have different emissivities. According to the *Handbook of Military Infrared Technology* (Office of Naval Research, 1965), emissivity is the ratio of the IR emitted by a surface to the IR that would be emitted by a blackbody at the same temperature.

A blackbody, which may not appear visually black, is a perfect absorber and emitter of IR, which means it has an emissivity of 1. Most surfaces aren't perfect blackbodies, and their emissivity is less. For this reason, most IR thermometers are set for a default emissivity of .95, which is close to the emissivity of most organic materials, including you and me. But shiny metallic surfaces, snow, brick, cotton fabric, and other materials may have lower emissivities. Some IR thermometers allow the emissivity setting to be adjusted so that temperature measurements are consistent.

Selecting an IR Thermometer

An online search will lead you to many different IR thermometers from a dozen or more manufacturers.

Some are intended mainly for measuring hot objects, while others have a much broader temperature response. Because I frequently measure the IR temperature of the sky, I prefer IR thermometers with the lowest temperature capability, which is around -76°F. Most IR thermometers have a minimum range of -20°F to -50°F.

IR thermometers are specified according to their field of view (FOV), which is expressed as a distance-to-spot ratio: the ratio of the distance to a target compared to the diameter of the spot viewed by the thermometer. For example, an FOV of 1:1 means the instrument will be looking at a circle 1 foot in diameter at a distance of 1 foot. A narrower FOV of 30:1 means it will be looking at a 1-foot circle at a distance of 30 feet.

For measuring very large objects, an FOV of from 1:1 to 10:1 should usually work. For smaller targets, select an instrument with an FOV of 30:1 or narrower.

Using an IR Thermometer

Be sure to read the instructions that come with an IR thermometer to learn how to operate the display backlight (if present), how to switch between Fahrenheit and Celsius temperature scales, and how to adjust the emissivity.

Keep in mind that IR thermometers are not designed to measure the temperature of shiny metal surfaces, which tend to have low emissivity. When pointed at objects on the far side of a glass window, they'll measure the temperature of the glass. Similarly, steam, smoke, and dust will also interfere with temperature measurements. Some manufacturers advise that their IR thermometers aren't intended for medical applications or for measuring body temperature.

Caution: Many IR thermometers include a laser pointer that illuminates the portion of an object that is being measured. Be careful to avoid pointing the thermometer at a person's face or at shiny surfaces that might reflect the laser beam into your eyes or someone else's.

Applications for IR Thermometers

There are dozens of applications for IR thermometers, some of which can provide very useful scientific information.

Studying Heat Islands

Heat islands are most often described as the dome of warm air that forms when sunlight heats buildings, streets, and other manmade things in cities. Actually, a heat island can be anything in the environment that becomes warmer than its surroundings, including rural roads, gravel, boulders, and remote farmhouses.

FIGURE 7-1. Paved sections of Hawaii's Mauna Loa Observatory road (left) are much warmer than nearby lava rock (right), presumably because of the lava's porosity.

You can sometimes see evidence of these heat islands, for example, when they cause snow around them to melt faster (see Chapter 2, "Snow Science"). The cedar elms on either side of my tiny rural office always leaf out faster than their neighbors due to the warmth of the building.

FIGURE 7-2. Four temperature readings, across and adjacent to Interstate 10 in West Texas on a clear spring day.

An IR thermometer is ideal for studying heat islands. Instead of simply assuming that an asphalt road is warmer than the grass that borders it, you can measure the exact temperature of both the road and the grass.

The temperatures of different building materials illuminated by full sunlight are of special interest. I've studied such materials by placing various bricks, concrete blocks, wood, and asphalt on a common background and measuring their temperatures over time. Also interesting are their rates of temperature decline when the sun is low in the sky.

FIGURE 7-3. Sarah Mims measured a newly paved asphalt road at nearly 225°F during a heat island study.

Studying Clouds

Both amateur and professional astronomers use IR detectors to detect clouds that might interfere with their telescope observations. An IR thermometer works well as a cloud detector because the base of a cloud is warmer than the clear sky.

Cumulus clouds form when the temperature of moist air falls to the dew point. The temperature indicated when an IR thermometer is pointed at the base of a low cumulus cloud is often within a few degrees of the dew

FIGURE 7-4. The open sky is always much cooler than the bases of cumulus clouds. The temperature straight overhead can be less than -75°F.

point measured at the Earth's surface. I've also learned that the temperature indicated when an IR thermometer is pointed at the clear zenith sky is well correlated with the total water vapor overhead. I'll soon have two years' worth of water vapor data and will write a scientific paper about this finding.

Energy Conservation and Storage

IR thermometers are very useful for evaluating the insulation in houses and other buildings. They're especially good at indicating temperature leakage through glass windows, and hot spots in electrical wiring.

An IR thermometer is also useful in developing solar water heaters. You can quickly try this method by placing a gallon of water in a black plastic trash bag. Seal the bag and lay it flat on a surface exposed to full sun. Measure the temperature of the bag every minute or two for 30 minutes and plot the data to see the increase in temperature over time.

Other Studies

Plants and animals provide many interesting temperature studies. For example, prickly pear cactus pads become much warmer than the leaves of trees and most other plants.

Fire ant hills can easily exceed 100°F on a warm day. I recently measured the surface of one at 114°F, then removed a 1-inch layer of soil from a portion of the hill, which then measured only 92°F. I wanted to continue

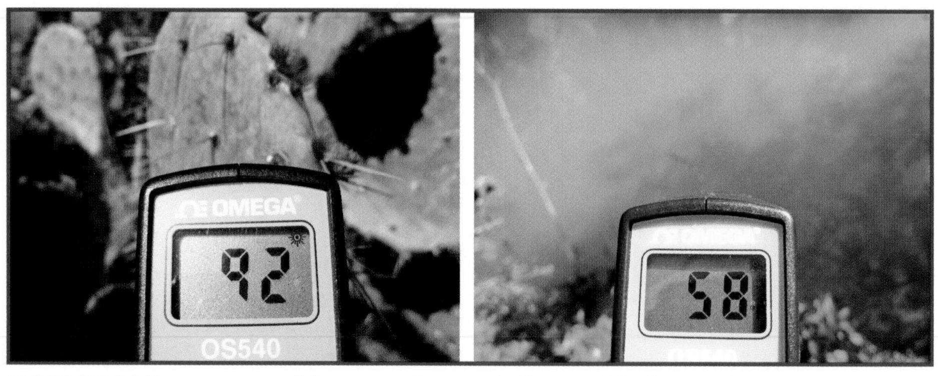

FIGURE 7-5. Prickly pear cactus pads become much warmer in direct sunlight than most leaves.

this investigation by going deeper, but hundreds of angry fire ants forced my retreat.

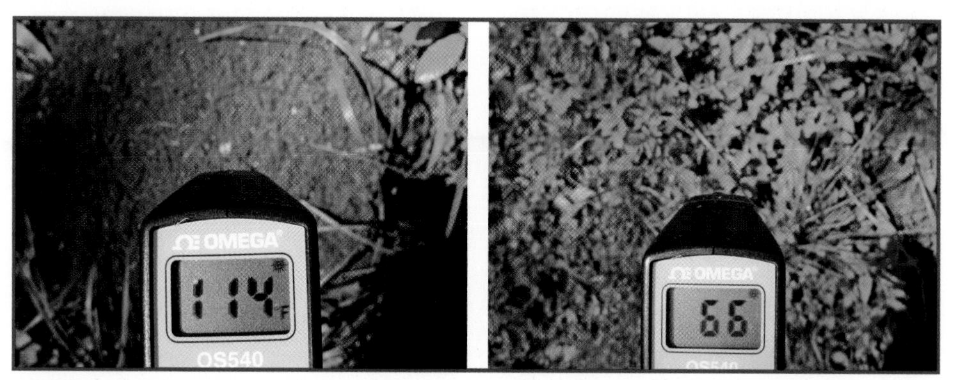

FIGURE 7-6. The temperature of this fire ant hill on a cool spring day was 114°F, while the adjacent ground was only 66°F.

Going Further

By now I hope you understand why an IR thermometer is one of the most important gadgets in my science kit. The ideas given here merely scratch the surface of what can be done with one of these remarkable devices, especially if you photograph the thermometer readout, as shown in the accompanying figures. Chances are you'll find entirely new uses when you begin probing your home, workplace, and environment with an IR thermometer.

8 Capturing and Studying Airborne Dust, Smoke, and Spores

Unless you're inside a laboratory clean room or wearing a dust mask, each breath of air you're now inhaling probably includes a fair number of dust particles and mold spores and, during spring and fall, a few pollen grains (Figure 8-1). If you're in good health, your respiratory system will filter out most of these particles.

You can find out more about these particles with an inexpensive microscope and a DIY air sampler.

FIGURE 8-1. Pollen blowing from a giant ragweed plant, one of the worst causes of hay fever.

Passive Air Samplers

In this chapter, we'll experiment with the simplest air samplers, those that rely on gravity or wind to deposit particles in the air onto adhesive tape or a bare microscope slide.

Let's begin with the microscope slide covered with ragweed pollen shown in Figure 8-2. Behind the slide is the windshield of my pickup truck, also coated with pollen. The pollen grains are held in place by the molecular attraction known as van der Waals force, which is strong enough to secure the pollen on the windshield of a car moving along a highway at 70mph.

Figure 8-3 shows an air sampler I've used while driving the highways of Texas and New Mexico. It's simply a plastic vacuum cleaner attachment with a microscope slide taped across the wide end. The assembly is taped to the hood or a side window of a car.

This passive sampler has collected a wide variety of particles in the air, including the Alternaria spore in Figure 8-4 that was photographed by holding a digital camera at the microscope eyepiece.

The biggest drawback of these passive-air sampling methods is that they provide only the particles, with no information about their concentration in the air. Yet even a simple passive air sampler can lead to a scientific discovery, as my daughter Sarah learned when she was in high school.

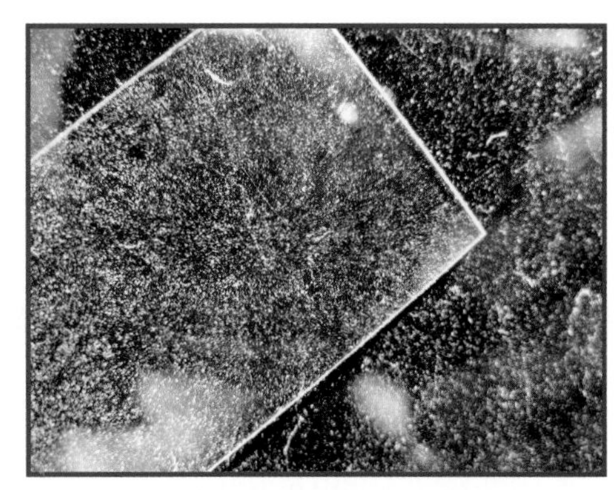

FIGURE 8-2. Pollen accumulates on windshields.

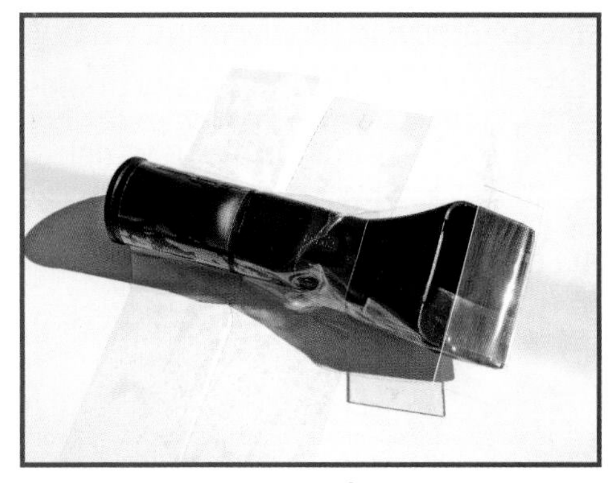

FIGURE 8-3. Air sampler made from a vacuum cleaner inlet fixture and a microscope slide.

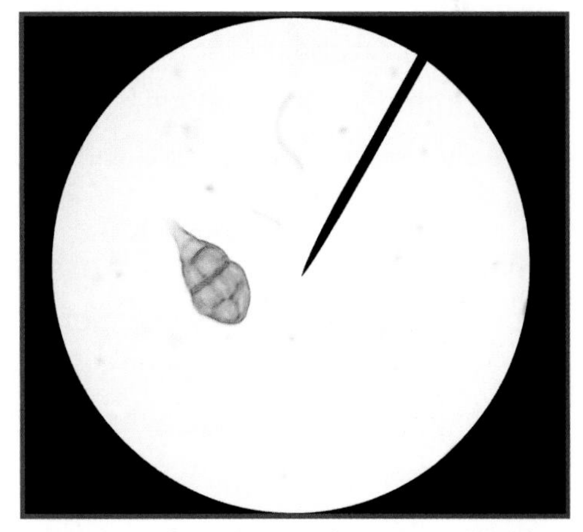

FIGURE 8-4. Alternaria spore, about 25μm, collected using the sampler. Ragweed pollen is about 20μm in diameter, and human hair 60μm-100μm.

"Stuff in the Air"

For her first major high school particle project, Sarah tried various methods to collect African dust that often drifts over Texas during summer. These included loops of bare adhesive tape, a car vacuum cleaner equipped with a filter, and a sampler she made from an enclosed muffin fan and a coffee filter. All these methods collected many dust particles, but they were difficult to find when the filters and tape were examined with a microscope.

Ordinary microscope slides worked much better. Each afternoon during her study, Sarah placed a clean microscope slide on an 11-foot tower in our field. She stored the exposed slides in a dust-tight microscope slide box that kept them separated.

She used online NASA satellite images and Navy Research Lab dust and smoke forecasts (www.nrlmry.navy.mil/aerosol/) to verify that the dust she collected fell from the sky on days when Sahara dust was at our place.

Sarah's Sahara dust project, which she called "Stuff in the Air," earned many science fair awards, including a grand prize from the Texas Junior Academy of Science.

"Smoke Bugs"

Sarah expanded her air sampling research to look for mold spores that might be accompanying dust from distant places. Scientists had already discovered that mold spores were crossing the Atlantic along with dust from Africa. So she decided to look for spores in dust that often arrives in the United States from Asia each spring. To do so, she exposed microscope slides each day for two weeks when Asian dust was expected over our place.

Although the Asian dust clouds passed north of our place, Sarah's microscope slides were loaded with spores. Atmospheric scientist Dr. Tom Gill told Sarah that smoke from agricultural fires in Yucatan, Mexico, was passing over our site during the week or so that she was exposing her slides.

Sarah used a microscope to find and count 22 genera of fungal spores and many black carbon particles (smoke) that she observed during four scans across the length of each microscope slide. She entered the data in a spreadsheet and made an x-y scatter graph of the number of spores versus the number of black carbon particles on each slide (Figure 8-5).

The correlation coefficient (r2) is a statistical parameter that expresses how well two variables agree with one another, with 1.0 being perfect agreement. In this case, r2 was 0.78, which strongly suggested that the fungal spores were associated with the smoke.

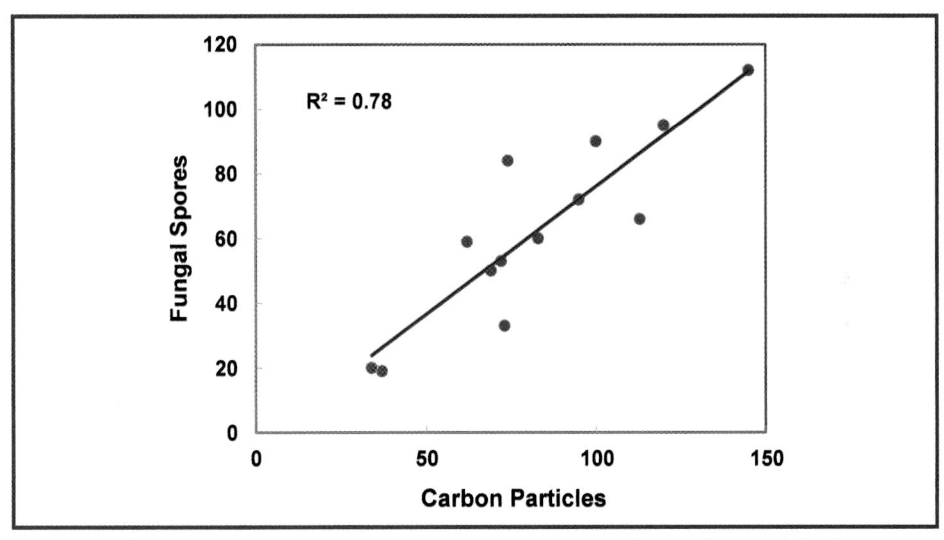

FIGURE 8-5. This scatter graph shows an association of mold spores and carbon particles (smoke) collected on microscope slides during 13 days when smoke from Yucatan drifted over South Texas.

Sampling the Air from a Kite

Sarah wanted to be sure that most of the spores on her microscope slide had arrived with smoke from Yucatan. The year after her initial discovery, we made two trips to the coast, the first when considerable smoke was arriving from Yucatan and the second several months later when there was little smoke from Yucatan.

On both days the breeze was arriving from Yucatan, and on both days Sarah stood in or near the water and flew a simple DIY air sampler from a kite at an elevation of about 50-100ft to avoid contamination from Texas spores (Figure 8-6 and Figure 8-7).

FIGURE 8-6. A DIY passive air sampler flies from a delta kite at Padre Island, Texas, to capture spores in smoke from agricultural fires in Yucatan.

FIGURE 8-7. At the bottom of the air sampler, binder clips hold the microscope slide in place.

After 30 minutes of exposure, she retrieved the kite, removed the sampler, and used her microscope to examine the slide (Figure 8-8). The results of the kite flights clearly showed significantly more spores and black carbon particles on the smoky day than on the clear day.

Sarah's discovery that smoke plumes from distant fires can include many spores and bacteria led to more awards, including another grand prize in the Texas Junior Academy of Science and a NASA web page (http://earthobservatory.nasa.gov/Features/SmokeSecret/).

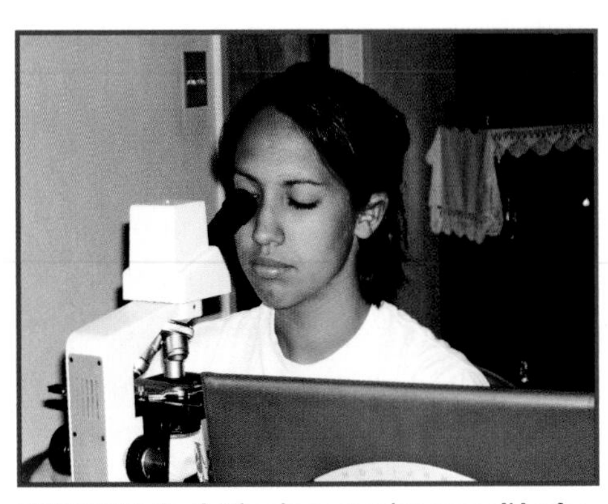

FIGURE 8-8. Sarah Mims inspects microscope slides for dust, smoke, and spores.

Since we could find no reports about the long-distance transport of microbes in smoke plumes, I helped Sarah convert her "Smoke Bugs" science report into a formal scientific paper, which was published on a priority basis by Atmospheric Environment ("Fungal spores are transported long distances in smoke from biomass fires," vol. 38, February 2004, pp. 651-655).

How to Make an Airborne Air Sampler

Smoke and dust can travel halfway around the Earth, and you can use the airborne sampler method to look for spores, carbon particles, pollen, and dust that arrive at your location.

Sarah made her airborne sampler from a 16oz plastic cup, two binder clips, a fishing tackle snap swivel, a microscope slide, and kite string. Follow these steps to make one in just a few minutes.

1. Use a ¼" paper hole punch to make 4 equally spaced holes below the rim of the cup (Figure 8-7).
2. Cut four 20" lengths of sturdy kite string and tie a string through each hole.
3. Tie the opposite ends of the 4 strings to the eyelet end of a sturdy snap swivel. (The snap end will be attached to the kite.)
4. Use a hobby knife to cut a 1" long slit, parallel to and 1" from the bottom of the cup. Use caution when making the cut.
5. Cut a ¼" slit from each end of the 1" slit straight down toward the bottom of the cup. Fold the rectangular cutout section outward.
6. Repeat Steps 4–5 on the opposite side of the cup.
7. Insert a clean microscope slide through one of the cutout openings so that its ends rest on the two cutout flaps.
8. Secure each end of the microscope slide to the plastic flaps with binder clips.
9. Protect the microscope slide from contamination by inserting the completed air sampler inside a clean plastic bag. Keep the 4 tether lines outside the bag, and secure the open end of the bag around the lines with a wire tie.

Flying the Sampler

The air sampler is designed to be flown from a kite rather than a balloon. The kite method assures that a strong current of air will flow through the plastic cup and deposit some of the particles it contains onto the microscope slide.

Sarah flew her sampler from a store-bought delta kite with a 33" keel and a 66" base (fully extended trailing edge). The kite proved surprisingly stable with the air sampler attached. Plan ahead and select a site that's well away from trees, power lines, and buildings. Avoid flying the kite if the wind is gusty or if dust is blowing nearby.

Just prior to releasing the kite, the protective plastic bag should be removed from the sampler. A flight of at least 10 minutes will probably be necessary. Keep the bag handy so you can store the sampler inside it after the flight.

Handle exposed slides by their edges. Store them in a microscope slide box with spacers, or cover the exposed side of the slide with clear adhesive tape.

Track the Leading Greenhouse Gas

9

For as little as $20, you can begin tracking the atmosphere's most important greenhouse gas, water vapor. And you can do so at any time, day or night, so long as the sky directly above you is cloud-free. This is a remarkable capability, especially since your measurements can have real scientific value if you make them at least once a day, weather permitting, from the same location.

Greenhouse Gases

The gases in the atmosphere that contribute to the warming of Earth are known as greenhouse gases. The best-known greenhouse gas is carbon dioxide, which plays crucial roles in the cycle of life. Every flower, fruit, tree, and animal is partially built from carbon extracted from the atmosphere. While carbon dioxide forms only about 0.039% of the atmosphere, it's increasing due to burning fossil fuels.

Water vapor is the king of the greenhouse gases, for it alone keeps the temperature of Earth warm enough to prevent the entire planet from freezing. Water vapor provides the precipitation that nourishes life, erodes rock, fills reservoirs, and provides hydroelectric power. The percentage of water vapor in the air varies dramatically depending on location, season, weather, and altitude.

The role of water vapor in regulating temperature becomes obvious when dry, cool air behind a cold front replaces moist, warm air. The same effect is observed when the air becomes both dryer and cooler when hiking or driving up a mountain.

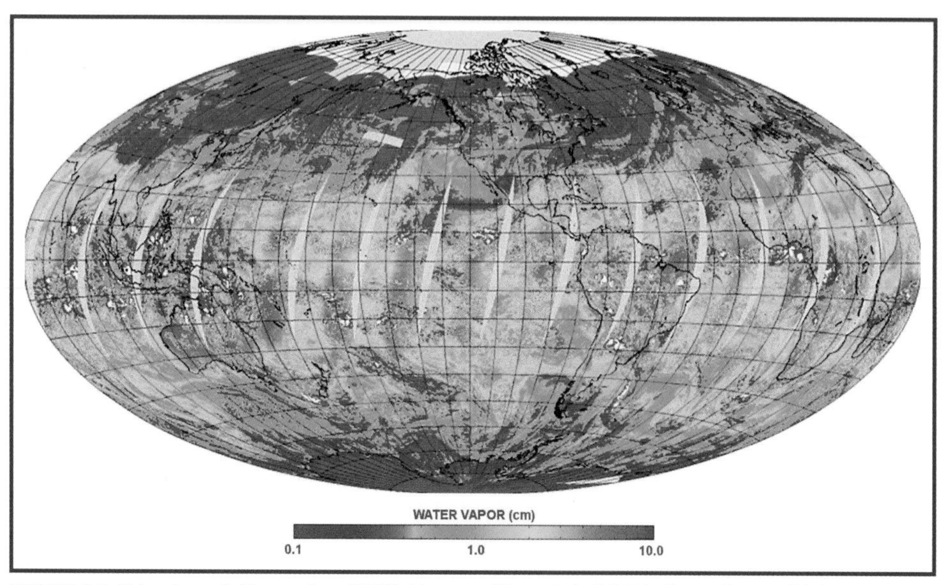

FIGURE 9-1. This color-coded image from NASA's Terra satellite on Feb. 4, 2011, shows the very high variability of total water vapor (PW) over Earth. The PW over a site in the U.S. can vary from a few millimeters after passage of a cold front to 6 centimeters or more on a warm summer day.

You can measure water vapor in the air around you with various kinds of humidity instruments. But these tools don't measure the total water vapor layer that is so important to the natural greenhouse effect. Meteorologists call this precipitable water (PW) or integrated PW (IPW), the equivalent depth of liquid water in a vertical column through the atmosphere.

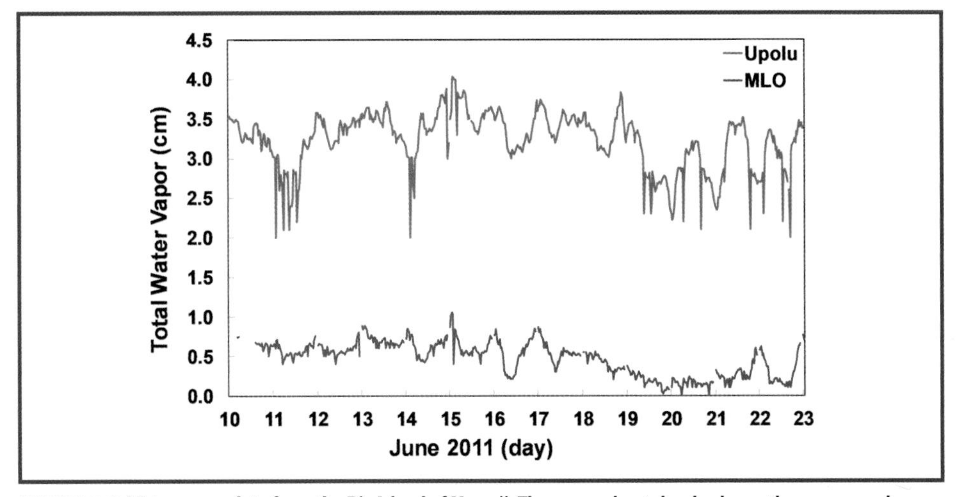

FIGURE 9-2. Water vapor data from the Big Island of Hawaii. The upper chart clearly shows there was much more water vapor over Upolu Point near sea level than over the nearby Mauna Loa Observatory, 11,200 feet above sea level, during the author's latest calibration visit. Data from NOAA's online GPS water vapor interface (http://gpsmet.noaa.gov/cgi-bin/gnuplots/rti.cgi).

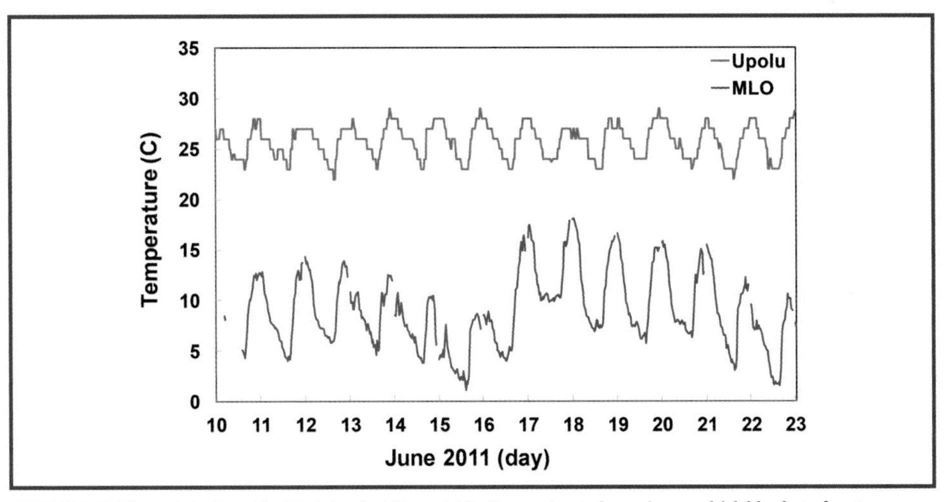

FIGURE 9-3. Vapor data from the Big Island of Hawaii. The lower chart shows how a thick blanket of water vapor keeps Upolu Point much warmer than MLO. Data from NOAA's online GPS water vapor interface (http://gpsmet.noaa.gov/cgi-bin/gnuplots/rti.cgi).

Weather balloons and a network of GPS receivers around the United States measure the total water vapor. Several satellites measure it from space, but they don't necessarily measure it through the entire atmosphere.

Join me and begin measuring PW, the layer of water vapor over your location. Since 1990 I've used an LED sun photometer to measure PW. For the past several years I've been using an even simpler method that anyone can use. It's as accurate as a sun photometer (about +/-10%) and it works day and night.

Using an Infrared Thermometer to Measure Total Water Vapor

Most of the visible sunlight arriving at Earth passes through the atmosphere and warms the water, soil, rocks, plants, roads, and buildings at the surface. All these warmed materials then emit infrared light, which allows their temperature to be measured with an infrared thermometer. Some of the infrared goes into the sky, where part of it is absorbed by water vapor and other greenhouse gases. These gases are warmed by the infrared they absorb. They then re-emit infrared, some of which is radiated back to the surface.

The same IR thermometer that can measure the temperature of the ground can be pointed straight up to measure the IR returned back to the surface by water vapor. Because of the thickness of the atmosphere and the wavelength response of an IR thermometer, the temperature indicated by an IR thermometer pointed at the sky is not necessarily the "sky temperature." Instead, the temperature indicated by the thermometer is directly proportional to the infrared emitted by the water vapor directly over your location.

Try to select an IR thermometer with a minimum temperature response of -60°C or less. I have used five different instruments, and best results have been provided by the IRT0401 and IRT0421, both made by Kintrex. The IRT0401 is not much larger than a lipstick tube, costs around $20, and looks at a field of view (FOV) of 53°. The IRT0421 costs under $50 and has a FOV of 5°. Both thermometers provide similar results, but the IRT0421 is best when clouds are near the zenith.

FIGURE 9-4. The inexpensive Kintrex IRT0401 infrared thermometer measuring the infrared radiation radiated down to the surface by the tiny amount of water vapor in the clear sky high over Hawaii's Mauna Loa Observatory. Sometimes the sky is so dry that there is too little radiation to be measured by these IR thermometers.

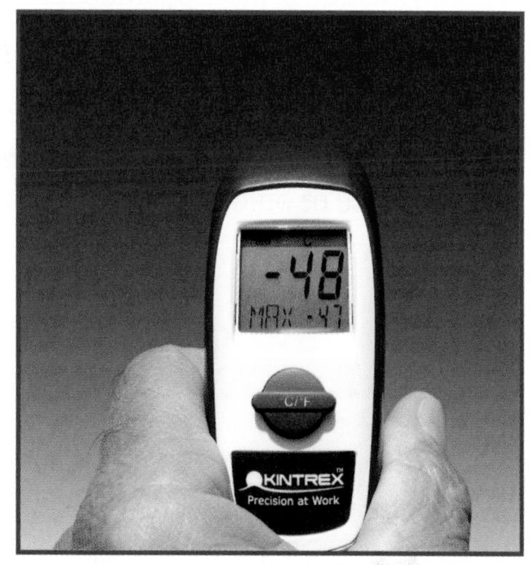

FIGURE 9-5. The inexpensive Kintrex IRT0421 infrared thermometer measuring the infrared radiation radiated down to the surface by the tiny amount of water vapor in the clear sky high over Hawaii's Mauna Loa Observatory.

Prepare to make a water vapor measurement by making sure the sky overhead is free of clouds. Face away from the sun, and hold the IR thermometer in your shadow so that the sensing aperture points straight up. Press the appropriate switch to make the measurement, and record it in a notebook along with the date and time. Be sure the thermometer doesn't see your head or hat while it's held in your shadow. During summer you may need to make measurements at mid-morning or mid-afternoon, when the sun is lower.

You will find that the sky "temperature" is considerably higher on moist days than on cold days. During very dry, cold winter days, your thermometer might not indicate a temperature. You can calibrate your thermometer to convert the temperature it indicates when pointed at the sky into the total water vapor over your head. The best way is to compare a series of measurements made over a range of dry and moist days with the total water vapor measured by the nearest GPS in NOAA's water vapor network. Make as many observations as possible before doing the calibration, including multiple measurements on days with rapidly changing conditions.

Some IR thermometers are equipped with an alignment laser. It's best to block its aperture with dark tape to prevent the beam from striking your eyes or those of onlookers.

How to Calibrate an IR Thermometer for Measuring PW

1. Make lots of sky measurements over several weeks, preferably during spring or fall, when weather is rapidly changing. You can use either the thermometer's Celsius or Fahrenheit scale so long as you're consistent. Fahrenheit provides better resolution.
2. Transfer your data to a spreadsheet program. Enter the date in column A, the time in column B, and IR temperature in column C.
3. Find the NOAA GPS IPW site nearest your location at http://gpsmet.noaa.gov/cgi-bin/gnuplots/rti.cgi. Download the data for the range of days you measured the sky temperature.
4. The GPS PW data are averages over 30-minute intervals. In column D, enter the PW measured closest in time to each of your sky temperature readings. The GPS data will be given in UTC, so be sure to convert your times to UTC when looking for the appropriate data. Add 5 hours to EST, 6 hours to CST, 7 hours to MST, 9 hours to PST, etc.

FIGURE 9-6. The author recently calibrated 7 infrared thermometers against the NOAA GPS water vapor receiver (white disk on pole) at Mauna Loa Observatory.

5. Use the spreadsheet to make an XY chart in which your temperature measurements are plotted on the x-axis and the GPS PW is plotted on the y-axis.
6. Use the spreadsheet regression feature to automatically fit an exponential curve to the points on the chart. Select the options for placing on the chart the equation representing the curve and R2, its coefficient of correlation (which indicates the goodness of the fit, with 1.0 being perfect agreement).
7. The equation for the best exponential fit to the data is the water vapor calibration formula for your IR thermometer. For most spreadsheets it will be of the form

   y = ex

   where x is an IR temperature measurement and y is the PW.

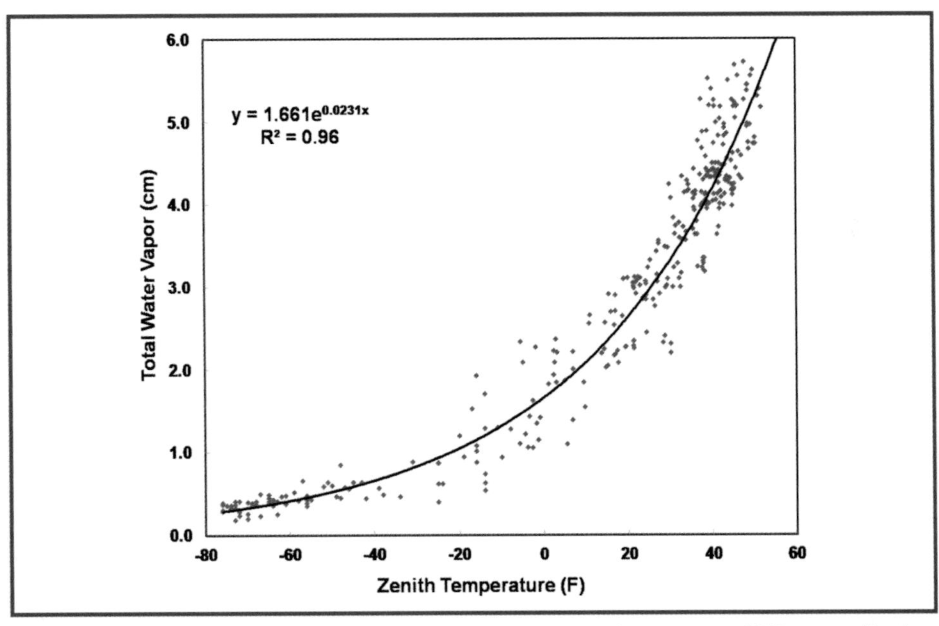

FIGURE 9-7. Scatter plot of sky temperature measured by an IRT0421 IR thermometer and PW measured by the nearest NOAA GPS from May to November 2010. The black line is the best fit to the data given by the equation at upper left. R2 is the correlation coefficient, and 0.96 indicates very good correlation..

For example, the exponential formula for one of my two IRT0421 IR thermometers is

$$y = 1.661e0.0231x$$

where y is the GPS PW in centimeters and x is the address of the spreadsheet cell containing the IR temperature of the sky. Enter the spreadsheet version of this formula into a single cell as:

```
=1.661*EXP(0.0231*C100)
```

where C100 is the cell containing x, the IR thermometer reading.

The spreadsheet will automatically do the math and place the PW amount in this cell.

You can copy the formula from your calibration into all the cells in column D of your spreadsheet. This will provide the PW each time you enter a sky temperature into the adjacent column C.

Build a Twilight Photometer, part 1

10

Have you wondered why some sunsets are so spectacular and others so drab? This ultra-sensitive photometer project will allow you to tease out the secrets of twilight, and even do serious science by finding the altitude of the dust, smoke, and air pollution that influence the colors of twilight.

With this project you can detect the tiny particles and droplets known as aerosols from 3km (around 10,000 feet) high to well above the top of the stratosphere at 50km (165,000 feet). While the photometer will not detect aerosols below 3km, many of those particles eventually float high enough to be detected. For example, from my Texas site I've measured the altitude of smoke from distant fires, haze caused by faraway power plants, and African dust that arrives every summer.

You can even measure the altitude of the sulfuric acid mist that forms an immense blanket 15km–30km high around our entire planet. This stratospheric aerosol layer, which major volcano eruptions can significantly enhance for several years or more, controls the duration of twilights and even influences climate.

Build a Simple Twilight Photometer

The twilight photometer shown in Figure 10-1 requires no optics and is considerably simpler, smaller, and cheaper than those used by professional scientists. Yet, as shown in Figure 10-2, it nicely estimates the altitude of dust and smoke clouds from 3km to 15km in the troposphere, and the permanent aerosol layer at around 15km–30km in the stratosphere. Instead of a conventional photodiode, the twilight glow is detected by an ordinary 660nm red LED or an 880nm near-infrared LED like those used in remote controls for TVs and appliances.

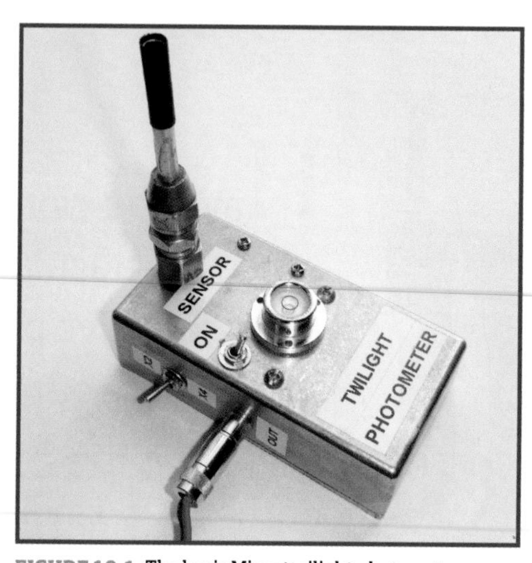

FIGURE 10-1. The basic Mims twilight photometer.

Materials

- Capacitors, ceramic: 47pF (1) and 0.01μF (1) designated C1 and C2, respectively
- Zener diode, 16V D1
- Operational amplifier (op-amp) IC, TLC271BIP IC1
- Clear capsule LED, 660nm red or 880nm IR
- LED socket (optional)
- Resistors: 40GΩ (2), 4.7kΩ (1). Use the 40-gigohm resistors for R1 and R2.

- Switches, miniature SPST S1 and S2
- Batteries, 9V
- Battery connector, 9V, with wire leads
- Battery support bracket (optional)
- Audio or phone plug and jack, 1/8"
- Output plug and jack compatible with data logger
- Insulated standoffs (2) RadioShack #2761381 or similar
- Perforated circuit board, 1 1/2"×1 ¼" with copper pads
- Metal enclosure
- Brass compression union fitting, 3/8"
- Aluminum or brass tube, about 4" length from hobby shop; to fit over your LED
- Bubble level
- Hookup wire and hardware

Tools

- Soldering iron and solder
- Drill and bits
- Multimeter
- Screwdriver
- Data logger (optional)

How It Works

The twilight glow straight overhead is very dim, and the photocurrent it generates in an LED is very small. Therefore, it's important to use an LED installed in clear epoxy. For best results, use an LED that projects a narrow

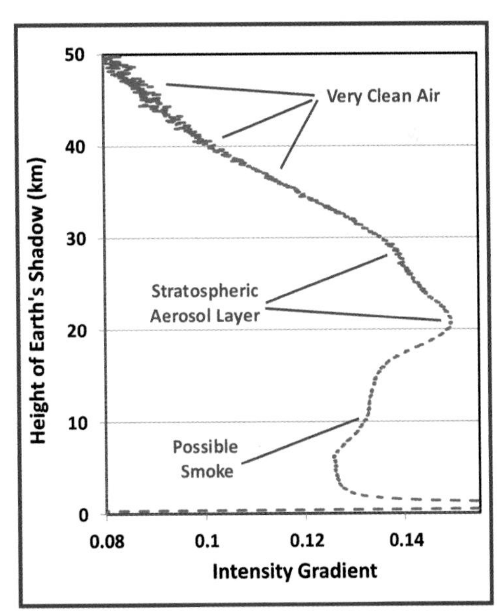

FIGURE 10-2. Measure the altitude of the sulfuric acid layer between 15km and 30km high.

beam when used as a light source. (For details about using LEDs to detect light, see Chapter 22, "How to Use LEDs to Detect Light.") The photometer circuit is shown in Figure 10-3. In operation, the tiny LED photocurrent is amplified billions of times, and transformed to a voltage by IC1, a TLC-271BIP operational amplifier with a very high-resistance feedback resistor consisting of R1 and R2 in series. Capacitor C1 suppresses oscillation.

The combined resistance of R1 and R2 controls the voltage gain of the amplifier. I have obtained best results using 40-gigohm resistors for both R1 and R2. When only 40 gigohms is required to provide a usable output signal during the 30–45 minute twilight period, switch S2 is closed to bypass R2. High-value resistors can be expensive and difficult to find, but I've had good results with Ohmite resistors from Mouser Electronics (www.mouser.com/) and Digi-Key (www.digikey.com/). For example, Mouser offers Ohmite's 40-gigohm axial-lead resistor (MOX-400224008K) for a reasonable $4.19. If 40-gigohm resistors aren't available, use 30- or 50-gigohms.

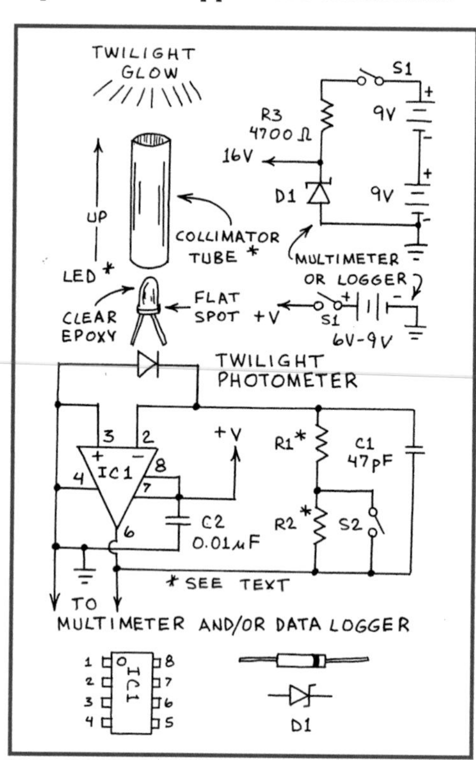

FIGURE 10-3. Holographic circuit schematic for the twilight photometer.

Planning the Photometer

The twilight photometer should be installed in a metal housing to block electrical noise from power lines and radio signals. I learned this lesson the hard way while testing my first twilight photometers at Hawaii's Mauna Loa Observatory. The LED can be installed inside the enclosure with a small hole to admit the twilight glow, or inside an open-ended phone or audio plug fitted with a collimator tube and inserted into a jack atop the enclosure. I've used both methods, and much prefer the external method described here for initial experiments. This allows you to try various LEDs and collimator lengths. After you find the optimum combination, you can install the system inside an enclosure.

Two 8-volt batteries connected in series power the photometer. Because IC1 must not be powered by more than 16 volts, the 18 volts from the two batteries is reduced to 16 volts by Zener diode D1. This arrangement provides the maximum possible output voltage range for a data logging multimeter. If you plan to use a DIY or commercial data logger instead (see Chapter 11, "Build A Twilight Photometer, part 2"), the circuit's output voltage must not exceed the data logger's allowable input voltage. This is typically 5 volts, which means you can power the photometer with a 5-volt battery instead of D1, R3, and the two 8-volt batteries. Both power options are shown in Figure 10-3. If the logger's input must never exceed 5 volts, insert a 1N914 diode between the positive battery terminal and IC1.

Assemble the Circuit

The circuit is built on a 1 1/2"×1 ¼" perforated board with copper traces on the bottom. The prototype board is shown in the open view of the photometer in Figure 10-4.

For best results, the op-amp's input (pin 2) should be isolated from the circuit board to prevent dust, fingerprints, and even the board itself from altering the gain of the op-amp. Isolating pin 2 in free air eliminates this problem. The easiest way to do this is to provide a 7-pin IC socket for IC1. Before inserting IC1 into the socket, bend pin 2 straight out so that it doesn't touch the socket when the other seven pins are inserted.

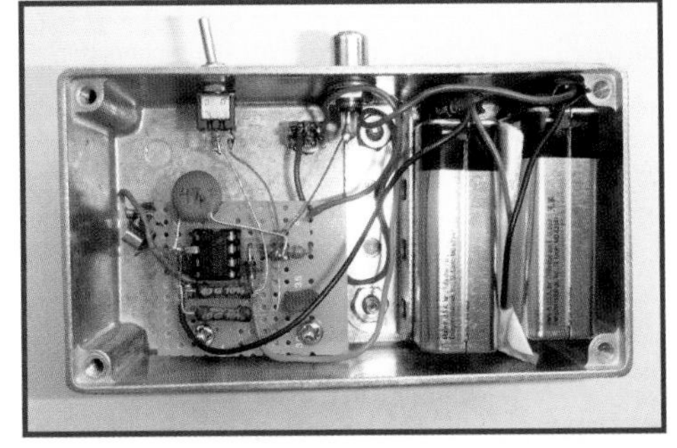

FIGURE 10-4. The twilight photometer is installed in a metal housing to block electrical noise from power lines and radio signals.

The next two steps are tricky, so refer to Figure 10-5 and take your time. First, solder the input side of R1 and C1 directly to pin 2. Then solder a wire directly between pin 2 and the LED cathode (-) terminal of the phone or audio jack.

Install the Circuit in an Enclosure

After the circuit board is assembled, clean the surfaces of R1, R2, and IC1 with a cotton swab dipped in alcohol. Then install the circuit board atop a pair of insulated standoffs inside a metal enclosure as shown in Figure 10-5. The photometer described here was installed in a Bud Industries CU-124 enclosure. You can use a larger enclosure or various metal containers from craft stores. If you use two 8-volt batteries (as shown in Figure 10-5), secure them in place with an angle bracket.

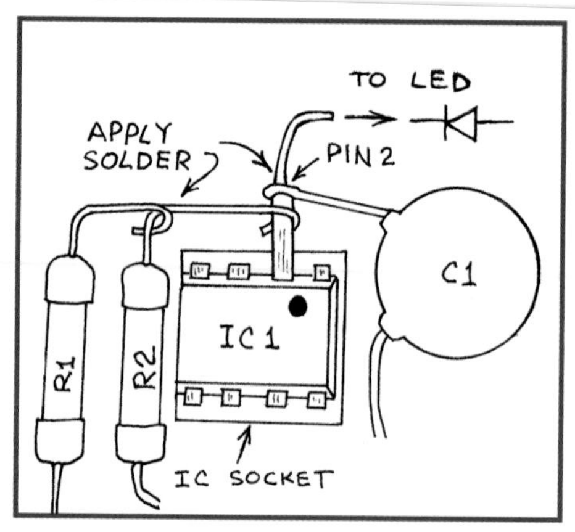

FIGURE 10-5. Closeup of the complex Pin2 schematic.

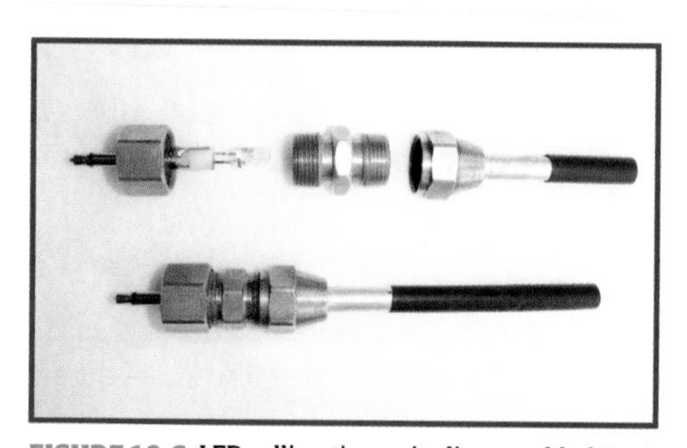

FIGURE 10-6. LED collimation unit, disassembled (top) and assembled (bottom).

Figure 10-6 shows a pair of LED-collimator assemblies made from a gas coupler fitting and an aluminum or brass tube. The LED is inserted into an LED socket (optional) soldered to 1/8" phone plug terminals or soldered directly to the phone plug. The phone plug is pushed into the open end of the lower coupler fitting and secured in place with a rubber O-ring.

You can omit the gas coupler if the plug's cap will slip over the LED, which means its leads must be clipped close and carefully soldered. Insert the open end of the plug into an appropriate collimator tube. A collimator length of 3"–4" should provide a field of view of around 5 degrees. As shown in

Figure 10-6, the length of the collimator can be increased or decreased with a short length of heat-shrink tubing.

Using the Photometer

On a clear day 10 minutes before sunset or 45 minutes before sunrise, place the photometer facing straight up on a level surface outdoors well away from light sources. A bubble level mounted atop the photometer will simplify alignment. If necessary, use shims to level the photometer. For best results, connect the photometer output to a data logging multimeter or a standalone DIY or commercial logger (Onset 15-bit HOBO UX120 or similar) and record data at 1-second intervals. If you don't have a logger, read the output voltage manually from a multimeter at 9- or 14-second intervals and enter the exact time and output voltage in a notebook or audio recorder. Automatic logging is preferable, but the manual method has been used for half a century.

Going Further

The raw twilight signal should provide a smooth curve when plotted on a graph of time versus signal. Far more significant is a graph that plots the rate of change in the data against the elevation of the twilight glow. Part 2 of this project, in the next chapter, explains how to process your data so that it accounts for these parameters and reveals the altitude of aerosols over your location. Meanwhile, you can learn much more about the science of twilight photometry at http://journals.ametsoc.org/doi/abs/10.1175/2008JTECHA1090.1.

Build a Twilight Photometer, part 2

11

Next time a Pinatubo-class volcano erupts, amateur scientists will be able to track the height of its aerosol cloud. Chapter 10, "Build A Twilight Photometer, part 1," showed how to make an ultra-sensitive DIY twilight photometer. In this chapter, you'll learn how to use a computer spreadsheet to manage and graph your photometer data so you can find the approximate altitude of layers of smoke, dust, smog, and volcanic aerosols in the atmosphere.

Tools

- Twilight photometer and bubble level (see Chapter 10, "Build A Twilight Photometer, part 1")
- Digital voltmeter
- Data logger (optional) recommended
- Compass or compass rose
- Computer with Internet access and LibreOffice spreadsheet software (free download at www.libreoffice.org)

The Twilight Glow

If you're looking toward the sun when it's just below the horizon, you are at the edge of Earth's shadow (Figure 11-1). You can see Earth's shadow for 10 or so minutes after sunset or before sunrise by looking opposite the sun. If the sky is clear, a pink band will form a wide arc over the horizon. This is the antitwilight glow. The gray or purplish sky below the arc is in the shadow of the Earth.

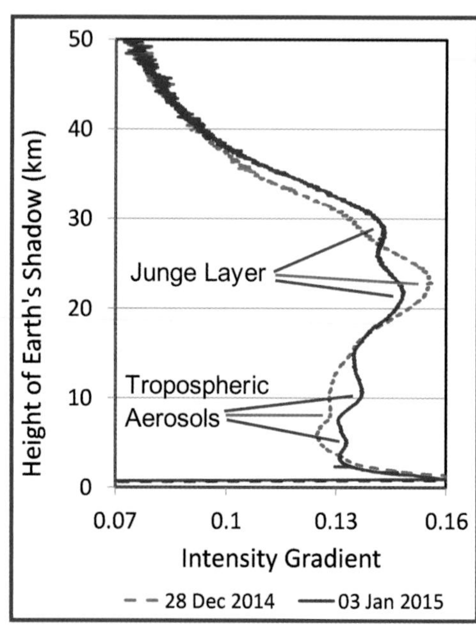

FIGURE 11-1. Height of Earth's shadow, and what you can see.

After sunset the antitwilight arc rises higher in the sky; the opposite occurs during sunrise. Because the atmosphere becomes less dense with altitude, the sky you see when looking straight up during this time is brightest just above Earth's shadow. Therefore, the intensity measured with a twilight photometer is highest just above the top of Earth's shadow.

This means the twilight intensity at any given time is approximately correlated with the height of Earth's shadow. If sufficiently dense aerosol layers are suspended in this region, the change they cause in the twilight signal can be detected and plotted.

Preparing the Twilight Photometer

It's important to adjust your twilight photometer to measure the widest possible range of twilight intensities. Ideally you'd do this by rotating a potentiometer shaft, but that's not feasible with our twilight photometer, because inserting a pot between the LED light sensor and the input to the op-amp might introduce noise. (And I'm unaware of inexpensive pots having a resistance of tens of gigaohms.)

The photometer includes two gain resistors, R1 and R2, connected in series, and switch S2 connected across R2. Closing S2 cuts the gain in half. This provides an X1 and X2 gain control. You can alter the gain even more by using different resistances, but this can become expensive.

A simple way to fine-tune the sensitivity of a twilight photometer is to alter the length of the collimator tube installed over the LED. Start with a 5"-6" length of heat-shrink tubing for the collimator tube, then increase the photometer's sensitivity by clipping off short segments of the tube to allow more light to enter it, until the photometer output is slightly below the maximum voltage allowed by your data logger. This should be done a few minutes before sunset. You'll need at least one twilight session to find the optimum length of the collimator tube. You can then replace the heat-shrink with a permanent piece of tubing.

Data Logger Selection

You can manually record your data at 14- to 30-second intervals, but automatic logging at 1-second intervals is much better. For this you'll need to connect the output of the twilight photometer to a 15-bit or higher resolution data logger. I've had good results with Onset's HOBO

UX120, a 16-bit, 3-channel analog logger (www.onsetcomp.com/products/data-loggers/ux120-006m). I've also used the Unisource DM620, a 50,000-count data logging digital multimeter. Many other data loggers and 50,000-count logger DMMs are also available. Just be sure the software is compatible with your computer.

The Twilight Photometer Spreadsheet

I've built a custom spreadsheet that manages and graphs your twilight data while saving you from having to solve many equations. You can download it for free and follow the detailed instructions for using it at the project page online: http://makezine.com/projects/twilight-photometer/. The spreadsheet was developed in Microsoft Excel and converted to free LibreOffice (www.libreoffice.org). It has 6 pages:

1. Analysis. This sheet calculates the times of sunset and sunrise, sun position, and height of Earth's shadow. It also calculates the derivative of the data (change over time), averages it (to smooth it), and creates the charts shown on sheet 2. References are provided to acknowledge those who devised twilight photometry.
2. Charts. This sheet shows graphs of the altitude of Earth's shadow versus the raw data (linear and logarithmic) and the derivative (intensity gradient) of the data.
3. Satellite. Satellite and aerosol forecast images are pasted here. Satellite images show any clouds that might be present. The aerosol models predict the distribution of dust, smoke, and smog.
4. Soundings. This is an optional sheet for upper air soundings from weather balloons launched closest to your site.
5. Data. Your raw data goes here.
6. Readme. Detailed photometer and data analysis instructions are here. Carefully read this sheet before your first twilight session.

Some Twilight Examples

Figure 11-2 shows the three graphs from two separate twilights superimposed, to demonstrate how the spreadsheet teases out the aerosol layers in the gradient graph. These graphs are from clear skies; irregular graphs are produced when clouds are present along the sun's azimuth.

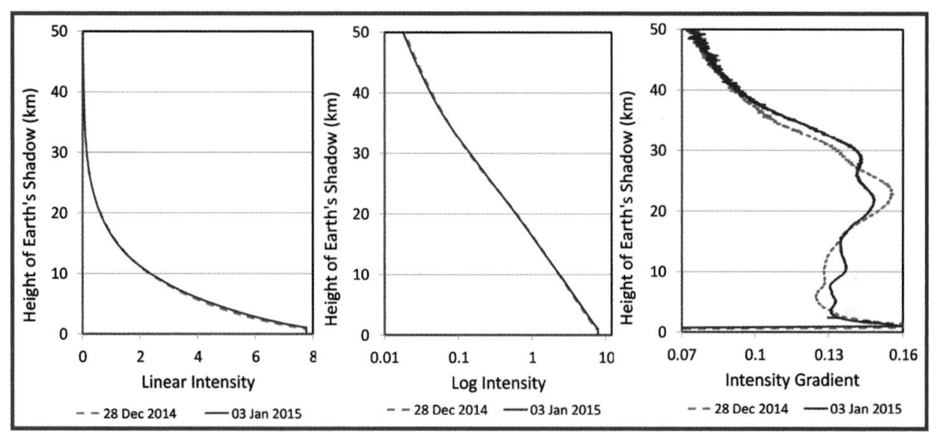

FIGURE 11-2. These graphs are from clear skies; irregular graphs are produced when clouds are present along the sun's azimuth.

Going Further

Twilight photometry is an ideal way to become better acquainted with the upper atmosphere. It's a useful tool for science fair projects, serious science, and curious sky watchers, especially if a major volcano eruption occurs. The twilight photometer spreadsheet also provides suggestions on how to expand the project and how to use a NASA aerosol model to identify aerosol layers you detect. When the next Pinatubo or Laki—or even a Villarica or Ontake or Eyjafjallajökull (if you live nearby)—blows its top, you'll be ready.

12 How to Analyze Scientific Images

Digital cameras are among the most important instruments in my science tool kit. They have provided thousands of images of twilight glows, solar aureoles, clouds, tree rings, insects, vegetation, bacteria and mold colonies, and much more.

Most images speak for themselves and need no more processing than that applied by your eyes and brain. But what if you want to extract data from images? Analytical studies of photographs require more than simply describing, say, the color of a leaf or the brightness of a sunset—they require numbers.

Prior to the digital era, scientists relied on instruments called densitometers to extract data from photographs. Transmission densitometers convert the degrees of darkness (or density) of points on photographic film into a representative voltage or signal. One side of the film is illuminated by a light source, and the light that leaks through is detected by a photosensor on the opposite side. Photographic prints are digitized by placing both the light source and the detector on the same side of the image.

While densitometry is still used to extract data from photographic film and prints, these can now be easily digitized by what amounts to a new kind of densitometer: the flatbed scanner. And the data in digital images is already present in a form that can be easily analyzed by image processing software.

ImageJ: Image-Processing Software That's Powerful, Platform-Independent, and Free

Processing digital images to extract their data once required powerful, expensive software. Its high price prevented many students and amateur scientists from analyzing their images.

This changed in 1997 with the introduction of ImageJ, a public-domain image analysis program developed by Wayne Rasband at the National Institutes of Health (NIH). During its first decade, ImageJ became a powerful, platform-independent image analysis package that can be run on Linux, Macintosh, and Windows machines. I've run it on a variety of Windows machines, including a tiny Acer Aspire One running Win XP with only 1GB of RAM (and a 160GB hard drive).

ImageJ requires no license, and the program and its Java source code are freely available at the ImageJ homepage: http://rsbweb.nih.gov/ij/index.html.

Running ImageJ

After you download the version of ImageJ designed for your operating system, go ahead and select and run the program. A rectangular command bar will appear at the top of your computer screen (Figure 12-1). This tiny but powerful startup menu is a toolbox that includes a set of 8 text selections over a row of 19 icons that point to menu options known as macros, some of which can be accessed immediately while others lead to still more macros.

FIGURE 12-1. The ImageJ menu window.

ImageJ's startup menu floats over the upper edge of whatever program was running when you clicked on it. You can use your mouse to drag it anywhere on your screen. If the toolbox disappears, simply click on the ImageJ icon on your task bar to place it back on your screen.

Doing Science with ImageJ

Let's get started with ImageJ by analyzing a photograph taken with a digital camera. The image in Figure 12-2 reveals a glow around the sun known as the solar aureole. The sun itself is blocked by the occluder device in Chapter 18, "How to Photograph the Solar Aureole."

FIGURE 12-2. Solar aureole image ready for processing in ImageJ.

First, select File > Open and choose an image, in this case the JPEG in Figure 12-2. After the image is displayed on the screen, maximize it. You're now ready to analyze the image.

This image was taken on a clear day at solar noon, and it can tell us much about the presence of aerosols in the sky, such as smoke, dust, and haze. You can quickly see how—simply place the cursor over various parts of the image and watch the five numbers that appear below the row of toolbar icons.

The first two numbers indicate the x-y coordinates of the position of the cursor on the image. The next 3 numbers indicate the intensity of the red, green, and blue (RGB) wavelengths of light at that position, on a scale from 0 to 255. Placing the cursor over the black occluder will cause each of the RGB numbers to fall below 20. Move the cursor over the nearby sky, and the blue number will increase much more than the red and green. The blue number will be highest when the cursor is moved all the way to the edge of the image.

Now it's time to let the program do all this for you, by drawing a graph of the intensity of blue light across the center of the image. First, make sure the ImageJ menu bar is displayed. Then click-drag with your mouse to draw a narrow rectangle across the entire image, centered over the black occluder. The rectangle will be outlined in yellow, and you can use your mouse to alter its size or nudge the entire rectangle into a different position. This process is called making a selection (Figure 12-3).

FIGURE 12-3. Selecting a region to be plotted.

Now you're ready to analyze the intensity of sunlight within your selection. Select Analyze > Plot Profile. A graph depicting the intensity of light across the rectangle you selected will almost instantly appear (Figure 12-4).

The solar aureole plotted in Figure 12-4 was for a clear sky. Figure 12-5 shows the plot for a day when a considerable amount of Saharan dust had blown from Africa to Texas. The images are clearly very different and so are the plots.

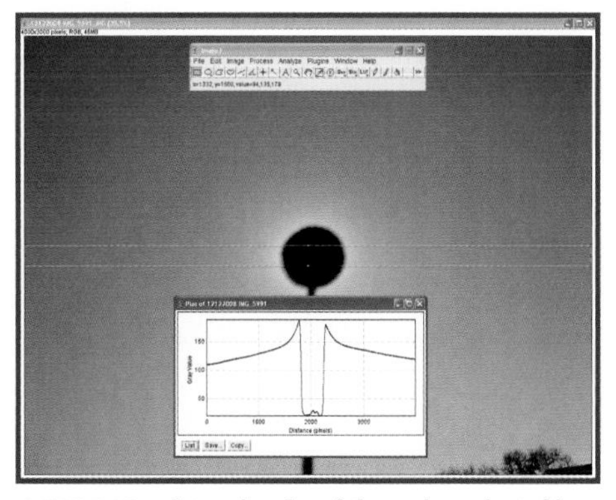

FIGURE 12-4. Intensity plot of the region selected in Figure 12-3.

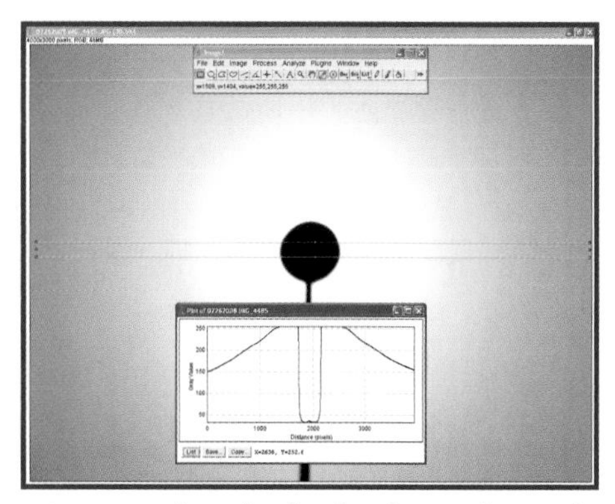

FIGURE 12-5. Intensity plot of a solar aureole caused by African dust over Texas.

The plots allow you to list the digitized values, which you can copy and paste into a spreadsheet like Excel or the free, open source OpenOffice Calc (www.openoffice.org). You can then make a chart like the one in Figure 12-6, which superimposes the data in Figure 12-4 and Figure 12-5 to clearly show the huge difference between the clear and dusty skies.

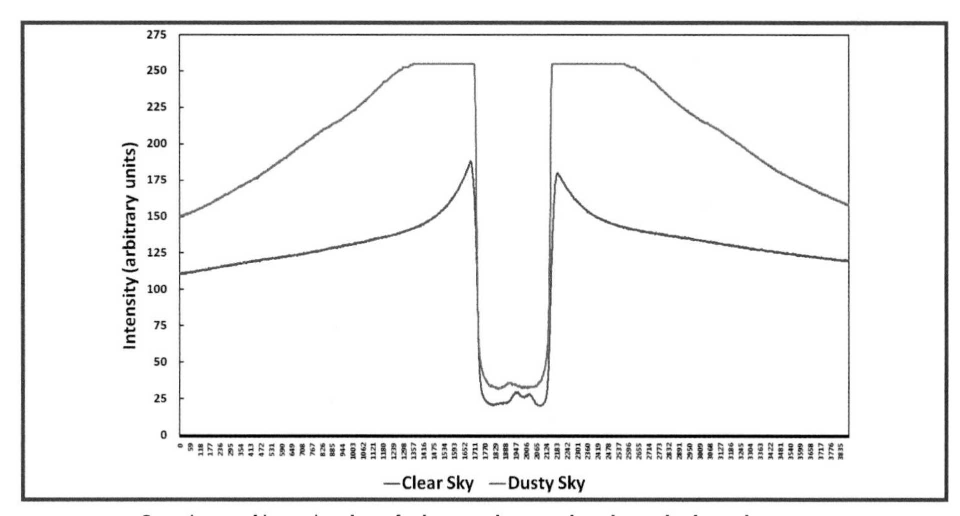

FIGURE 12-6. Superimposed intensity plots of solar aureoles on a clear day and a dusty day.

Going Further

We've barely begun to exploit the capabilities of ImageJ with these simple examples. Want to see the distribution of separate red, green, and blue wavelengths? Select Image > Color > Split Channels. The program will quickly present you with three grayscale images, for R, G, and B. You can then analyze each image any way you like. For example, Figure 12-7 and Figure 12-8 show 3D surface plots for the red wavelengths of both the clear and dusty aureoles above.

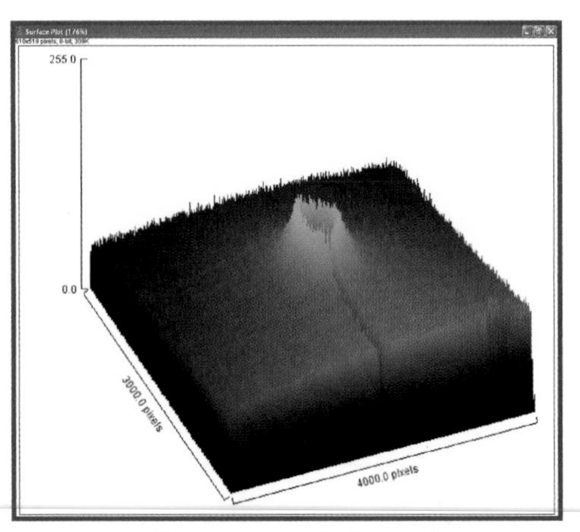

FIGURE 12-7. 3D surface plot of solar aureole, in a clear sky.

You can learn much more about ImageJ from the website, but the best way is to select some nice images, download the software, and take the plunge. You'll be amazed by this program's capabilities. Although it's fun to explore what ImageJ can do, you can use it for serious science. For example, image analysis software was the only way I was able to write a scientific paper based on an analysis of solar aureole images made with an old 1998 Fuji with only 1.3 megapixels resolution (F.M. Mims III, "Solar aureoles caused by dust, smoke, and haze." *Applied Optics* 42: 492-496, 2003).

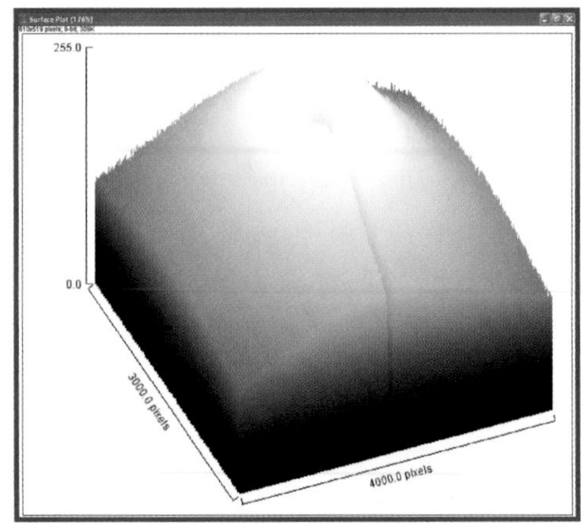

FIGURE 12-8. 3D surface plot of solar aureole, in a dusty sky.

Of course, ImageJ isn't limited to analyzing sky photos. One of my favorite activities is studying tree rings. In the next chapter, we'll see how ImageJ can transform this hobby into serious science, too.

13 Doing Science with a Digital Scanner

The transition from film to digital cameras has made a huge impact on how amateur scientists can save, analyze, catalog, and publish their imagery. It's safe to say that digital cameras and personal computers are among the most important tools in the amateur scientist's kit.

When the subject is two-dimensional, flatbed digital scanners can also play a major role in imaging science. They are ideal for making high-resolution images of leaves, dragonflies, butterflies, tree ring sections, soil samples, and many other subjects.

Advantages and Disadvantages of Scanning

Virtually shadow-free lighting is a key advantage of a digital scanner—the scanner provides its own light source. Another advantage is that scanners don't suffer from the distortions caused by camera lenses. Scanners are relatively inexpensive, and they can be used for many applications beyond the scientific roles described here.

Besides their two-dimension limitation, a major drawback of scanners is that objects being scanned must fit within the scanner's image plane. Scanners are also much larger than digital cameras.

Background Color

Most objects I've scanned looked best with a white background. Because my scanner (HP Scanjet 3970) has a 35mm slide scanner slot built into its lid, it's necessary to cover the object being scanned with an uninterrupted background. Two or three sheets of white 20-pound paper work well.

Light-colored objects are not easily visible when scanned against the white background of a typical scanner lid. To provide contrast with light-colored objects, place black construction paper over the object being scanned before closing the lid. You can use various colors for special effects.

How to Scan Dragonflies

For years I've photographed dragonflies and damselflies by quietly sneaking up behind them. With practice, it's possible to get within a few inches of some species. Entomologists Forrest Mitchell and James L. Lasswell of the Texas AgriLife Research and Extension Center at Stephenville, Texas, have used digital scanning to create an impressive library of top and side views of many dragonflies and damselflies. You can see these images online at Digital Dragonflies (http://agrilife.org/dragonfly/).

The images at Digital Dragonflies were made by cooling each specimen in a refrigerator to keep it still while it was being scanned. The insects were then placed backside down on the scanner's glass bed. They were protected from being crushed by the scanner's lid by placing them inside a 10cm × 12cm rectangle cut in a mouse pad (see their website for details). The mouse pad approach is probably best, but I've found that ordinary corrugated fiberboard will also work.

FIGURE 13-1. This living dragonfly was scanned by placing it inside an opening cut in a sheet of corrugated board.

Based on my experience scanning dragonflies (Figure 13-1), butterflies and moths could also be scanned. A severe drought has slashed the butterfly population in Central Texas: I'll try this method as soon as the butterflies return.

Fossils and Artifacts

Fossils and artifacts having a flat surface are easy to scan. For example, I've scanned the fossil of a trilobite with good results (Figure 13-2). Even ripple marks in the mud (now shale) on which the creature was resting were captured.

Years ago I found a flint artifact in the gravel bottom of the creek that borders our land (Figure 13-3). You can also scan flint arrowheads and spear points.

FIGURE 13-2. Many fossils and flint implements are flat enough to permit scanning.

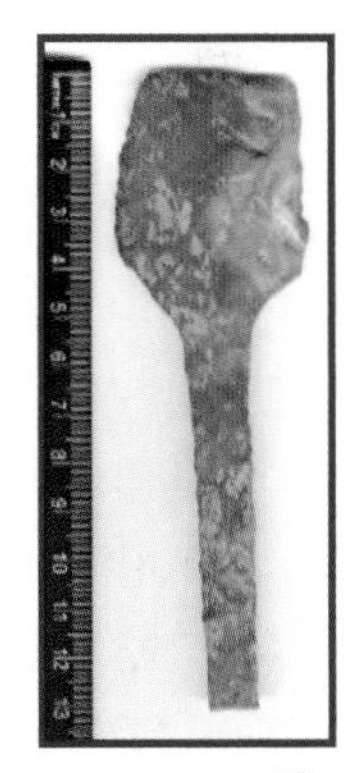

FIGURE 13-3. The artifact is a flint scraper that is very flat on both sides, easily scanned against a white background.

Plants

It's easy to scan the leaves of many kinds of plants against a white background (Figure 13-4). The main limitation of scanning leaves is the size restriction posed by the scanner's bed.

For best results, scan plant samples as soon as possible after collecting them. If this isn't possible, preserve the sample in a cool location or immerse its stem in water. Leaves may be coated with wax, so be sure to clean the scanner's glass bed after scanning.

FIGURE 13-4. Be sure to clean the glass surface of your scanner after scanning plant leaves!

The color of scanned leaves may not appear true to life. If not, you can use a photo-processing program to correct the color. I've found that the best way to accomplish this is to hold the actual sample next to the monitor and adjust the color of the image until it matches that of the sample.

Feathers

Most flight and tail feathers are easy to scan (Figure 13-5), but white feathers and those with white down require a dark background. For example, I once scanned a flamingo feather. The upper half of the feather was pink, and it scanned fine. Because the lower portion of the feather was white, it was barely visible against the white background. The entire feather was visible when a sheet of black paper was placed over it.

FIGURE 13-5. Feathers, being essentially flat, are easy to scan.

Soil and Sand

FIGURE 13-6. This piece of ancient amber was sliced and scanned to study the sand it captured when it flowed from a tree in what is now the Dominican Republic.

Soil and sand samples are easy to scan. Light-colored samples require a black background. Scans like this are important for soil science, since they allow different specimens to be compared under identical lighting conditions.

As with all scans, you might need to adjust the colors of scanned soil and sand samples. For example, when I scanned samples of light-colored sand from three locations (Figure 13-7), the sand that looked whitest to the unaided eye was not as white when scanned.

FIGURE 13-7. Samples of light-colored sand were scanned against a black background. From L to R, Padre Island, White Sands, Florida Panhandle.

Water can cause the color of soil and sand to change appreciably, and a scanner lets you scan both wet and dry specimens simultaneously.

Tree Rings

In Chapter 1, "How To Study Tree Rings," I described how to scan the annual growth rings in tree trunks and branches so they can be analyzed with ImageJ photo analysis software. Briefly, use a sharp, fine-toothed wood saw to cut sections from branches and trunks. Sand the exposed face of a section with 100-grit paper followed by 220-grit. The final polish is made with 400- or 600-grit paper (Figure 13-8).

Let wood dry for at least a day before sanding if it was cut green dry. You can enhance the appearance of the rings by moistening the face of the sample before scanning.

FIGURE 13-8. Growth rings in tree branches and small-diameter trunks are easy to scan. Shown here is a cross-section of a fir branch.

Microbe Cultures

You can easily save bacteria and mold cultures by digital scanning. My daughter Sarah's discovery of living microbes in biomass smoke arriving in Texas from the Yucatan Peninsula was made possible by exposing Petrifilms (made by 3M) to ambient air on days with and without smoke.

(See the article: Sarah A. Mims and Forrest M. Mims III, "Fungal spores are transported long distances in smoke from biomass fires." *Atmospheric Environment* 38: 651-655, 2004. http://earthobservatory.nasa.gov/Features/SmokeSecret/smoke_secret2.php)

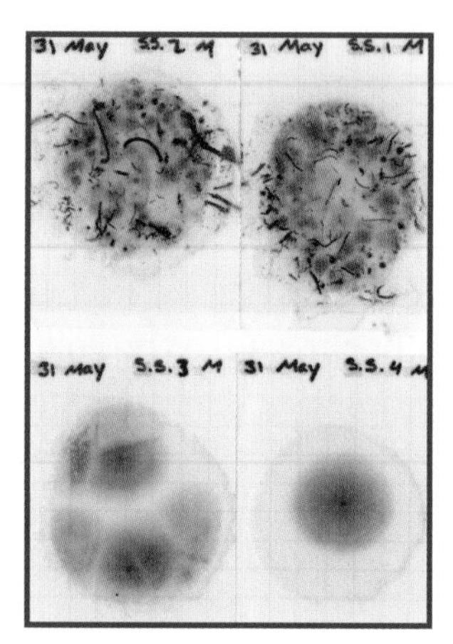

FIGURE 13-9. Colonies of bacteria and fungus spores can be easily scanned when they are grown on Petrifilm plates. These two films both show fungus colonies. The upper film was exposed to smoke from burning grass, and the lower film was exposed to nearby clean air.

To confirm her discovery, Sarah exposed Petrifilms to smoke from burning grass and nearby clean air. The films exposed to clean air had only a few bacteria and mold colonies. Those exposed to smoke had dozens of colonies—plus black ash from the burnt grass. Scans of these films provided persuasive evidence of the presence of microbes in biomass smoke (Figure 13-9).

Notes, Sketches, and Drawings

It's easy to scan field notes and sketches for posting on the web or sending to friends, or to simply provide backup.

My ambition is to scan all of my field trip notebooks, photographic slides, prints, and the thousands of log sheets on which I've recorded sun and sky data since May 1988. This will be a major chore that will require at least 10GB of storage. But the entire collection will fit on a tiny flash drive instead of a couple of heavy file cabinet drawers.

Going Further

Some specimens benefit from the inclusion of a scale in the scanned image. If the scale isn't perfectly straight, you can straighten the image in a photo-processing program.

You can use free, powerful ImageJ software (download it at rsbweb.nih.gov/ij) to analyze your scanned images. (See Chapter 12, "How to Analyze Scientific Images," for details about using ImageJ.)

Finally, there are many non-science applications for digital scanners that go well beyond making copies of documents and photos. For example, collectors of stamps and coins can easily digitize their collections.

14

Data Mining: *How to Analyze Online Scientific Data*

The Internet holds a treasure trove of scientific data. Never before has it been so easy for students, amateur scientists, retired professionals, and anyone with reasonable analytical and computing skills to do serious science. Even people without field or bench experience who have never used, much less built, an instrument can now make discoveries, possibly significant ones.

Data Resources

No matter what interests you most in science, you can probably find data that meets your interest somewhere on the web. Considerable online data is in the form of time series, collected at intervals ranging from seconds to years.

Here are a few sources:

- Tree ring data: www.ncdc.noaa.gov/data-access/paleoclimatology-data/datasets/tree-ring
- Ozone layer: http://ozoneaq.gsfc.nasa.gov/
- Weather (U.S.): http://ngdc.noaa.gov/
- Sunspots: www.sidc.be/silso/datafiles
- Cosmic rays: www.ngdc.noaa.gov/nndc/struts/results?t=102827&s=1&d=8,270,9
- Sea level change: http://sealevel.colorado.edu/
- Stream flow (U.S.): http://waterdata.usgs.gov/nwis/sw
- Satellite: begin your search at www.nasa.gov/about/sites/index.html

Software Tools

While a huge array of software is available for analyzing data and making charts, you can achieve much with a spreadsheet program. Excellent alternatives to commercial spreadsheet programs are OpenOffice (www.openoffice.org) and LibreOffice (www.libreoffice.org), a freeware package that includes Calc, a spreadsheet program with many of the features of Excel.

A Data Mining Example

Since Feb. 4, 1990, I've made nearly daily measurements of the ozone layer, total column water vapor, aerosol optical depth (a measure of haze), solar UV-B, and other atmospheric parameters from east of San Antonio, Texas. Now that this time series is 20 years long, I've become very interested in finding other long series of data from my area. The National Weather Service is one of many sites I've mined.

Formal temperature measurements at San Antonio began in 1885, and Figure 14-1 shows a chart of the average temperature for each year. The warming era of the 1930s, the present warming, and the intervening cooling are especially obvious.

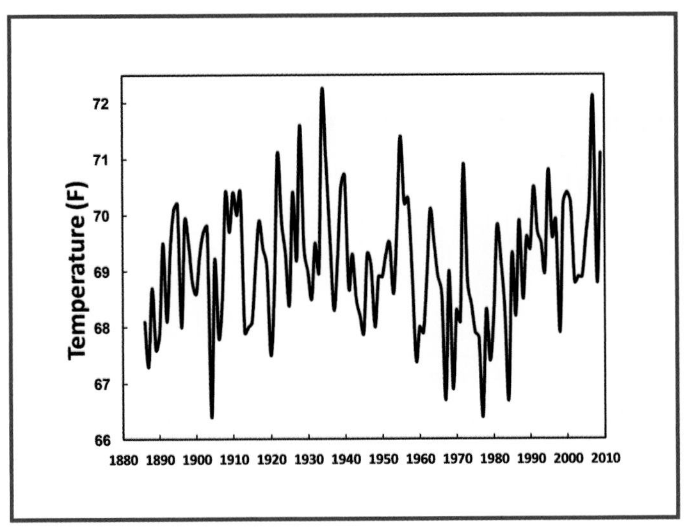

FIGURE 14-1. Preliminary annual temperature measured by the National Weather Service at San Antonio, Texas, from 1885 to 2009.

To make this chart, the data was saved as a text file and imported into Excel. The temperature data were given in degrees Fahrenheit, even though most countries and science publications use Celsius.

An advantage of the Fahrenheit scale is that it provides better resolution than the Celsius scale. You can easily convert Fahrenheit data to Celsius using the standard formula. The Excel version is

```
=(5/9)*(T-32)
```

where T is replaced by the column and row (e.g., P18) of the cell containing a temperature in Fahrenheit.

Figure 14-2 shows how the basic time series chart in Figure 14-1 can be embellished with a 9-year running average superimposed over the annual data. This addition serves to smooth the data and reveals patterns over time.

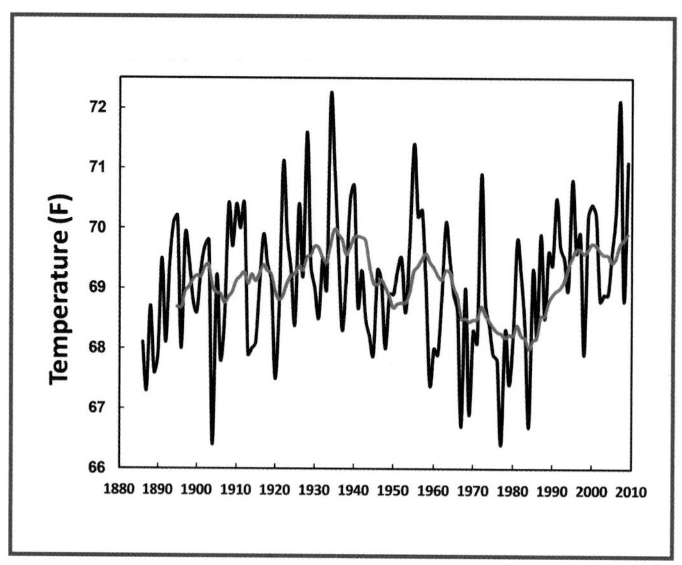

FIGURE 14-2. Figure 14-1 with a superimposed 9-year running average that more clearly reveals patterns and trends in the data.

Figure 14-3 replaces the running average with a regression line that shows the linear trend of the entire data series. A line representing the mean of the data has been added so that the upward slope of the regression line can be easily visualized without cluttering the chart with grid lines.

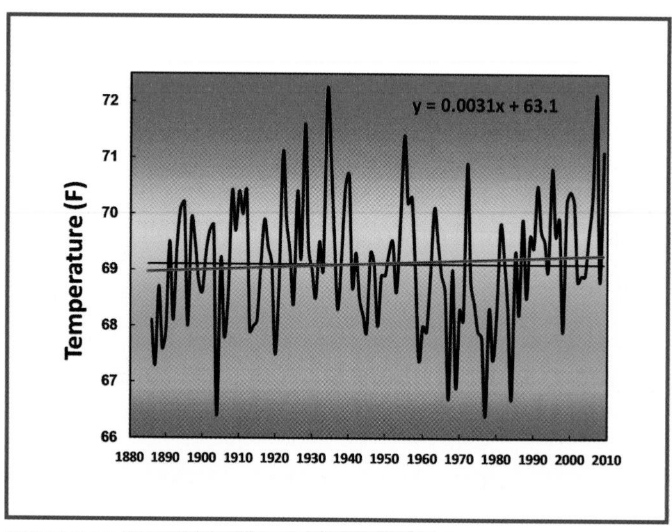

FIGURE 14-3. Figure 14-1 with a superimposed mean and linear trend (given by the equation at top) and a color-coded background.

Note the equation for the trend line in Figure 14-3. This establishes the slope of the trend and allows you to determine how much the temperature has changed from the starting point to any year.

The regression equation is

$$y = 0.0031x+63.1$$

where y is the temperature (T) and x is the year. This formula gives T = 68.94 in 1885 and 69.33 in 2009, for an increase of 0.38 degrees.

Figure 14-3 is decorated with Excel's *gradient fill* option to provide a hot/cold (red/blue) color-coded background for the chart. This looks good on the web and in general publications, but would be inappropriate in a formal, peer-reviewed paper in scientific journals.

While it's common for climate scientists to plot linear trends of their data, this method misses significant fluctuations in the data.

For example, the trend lines in Figure 14-2 and Figure 14-3 completely miss the warm temperatures of the 1930s and the cool temperatures of the 1970s. This and the uncertainty of many kinds of experimental data mean a linear trend line cannot always forecast changes to come.

Some Caveats

It's all too easy for statistics and pretty graphs to overshadow factors that might have influenced the outcome. For example, how reliable is the warming trend depicted in Figure 14-1, Figure 14-2, and Figure 14-3? The NWS website for the San Antonio temperature data states, "Please note that these data are preliminary and have not undergone final QC [quality control] by NCDC [National Climatic Data Center]. Therefore, these data are subject to revision." This caveat is important for multiple reasons, since some NCDC adjustments to the climate record have been questioned.

Then there's the urban heat island effect. Based on a study at nearby New Braunfels, the urbanization of San Antonio has probably caused that city's temperature to rise several degrees since 1885. Thus, the increase of 0.4° F shown in Figure 14-2 and Figure 14-3 might amount to a decline of several degrees in the nearby country. These caveats are amplified by the fact that the San Antonio weather station was moved several times since 1885.

Don't let these caveats discourage you. Just do your best to evaluate anything that might have affected the data you are studying. After all, even professionals face the same kinds of uncertainties.

Going Further

Figure 14-4 shows a solar ultraviolet-B radiometer on a rooftop at Texas Lutheran University in Seguin, Texas. This is part of a suite of U.S. Department of Agriculture sunlight instruments that I have managed for Colorado State University since 2004. It's one of more than 30 such instruments across the United States, and all the data are freely available at http://uvb.nrel.colostate.edu/UVB/index.jsf.

Figure 14-5 is a work in progress, a chart of 87,977 measurements of erythemal (sunburning) UV-B made by the radiometer in Figure 14-4 every three minutes during 2009. The feature in red is a one-week running mean of the data that clearly shows changes caused by the seasons and cloudy periods. This is one of many such charts I'm using to explore the data.

FIGURE 14-4. UV-B data from this U.S. Department of Agriculture radiometer at Texas Lutheran University at Seguin, Texas, are online.

With patience and care, you, too, can make and analyze plots of data available online. But this doesn't mean you should rush your findings into print or onto the web. Posting or publishing research findings requires considerable care to avoid making mistakes and drawing the wrong conclusions. So be skeptical of your results and move forward with care. Learn basic statistics, explore your spreadsheet's functions, and find out how to add error bars to your charts.

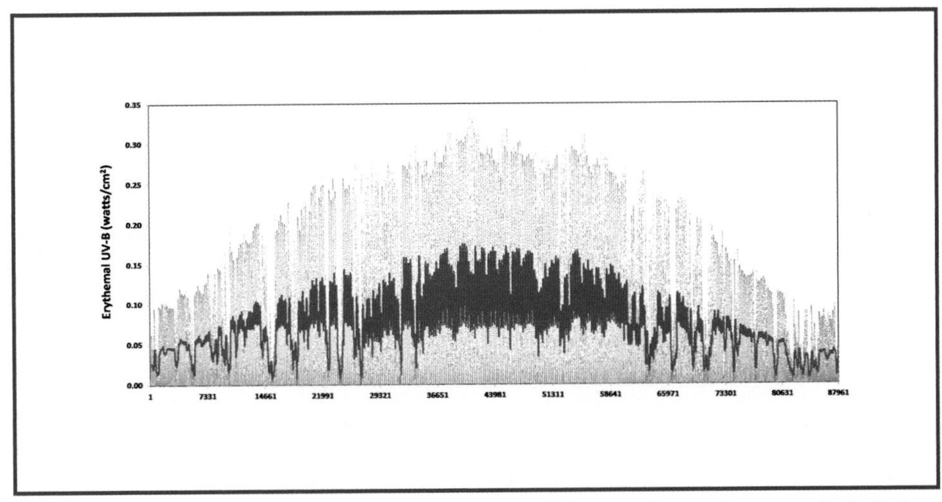

FIGURE 14-5. The 87,977 readings (blue) every three minutes and the seven-day running average (red) of solar ultraviolet radiation during 2009 measured by the USDA instrument in Figure 14-4. Because of a few days of data outages, it wasn't possible to label the y-axis with dates. That's why I call this chart a "work in progress."

If you're investigating an area of science that's new to you, be sure to seek advice from objective professionals. If you can demonstrate that you're serious about your research and want to do it properly, chances are they'll provide the advice you need.

If you do post or print mined data that you've analyzed, it's important to acknowledge those who collected the data and the website or publication where you found it. As a courtesy, you might want to first show your analysis to those who collected the data. They might be willing to check your work for errors and even establish a relationship for future collaborations.

15 Using Sensors with Dataloggers

Data logging is the automatic collection and saving of information. Having the ability to automatically log a string of measurements and save them for later study and analysis can help transform you from an experimenter into a scientist.

Electronic data logging provides an enormous range of opportunities for amateur scientists. Back in the 1980s, I would connect a sensor to the joystick port of a PC like RadioShack's Color Computer or IBM's PCjr and write simple data logging programs in BASIC. When analog-to-digital interface boards became affordable, my logging became more sophisticated. Still, these methods required a computer dedicated solely to the logging operation.

Data logging changed dramatically when miniature, standalone loggers were introduced. These devices provide real-time logging without the need for a dedicated computer. They're activated and downloaded by a computer, but in between, they operate independently. Data loggers are available from Onset Computer, Jameco, Omega Engineering, SparkFun, and others. Maker Shed (www.makershed.com) also sells a data logging shield for Arduino microcontrollers.

Interfacing Sensors to Loggers

You can buy tiny data loggers that store temperature readings over time. Other loggers record light, carbon dioxide, pressure, and other parameters. But what if you want to log a parameter for which there is no logger?

You can build your own logging system from scratch or from published plans. Or you can do as I've done and design DIY sensor circuits that can be connected to the voltage inputs of commercial loggers. This approach frees up considerable time, since software is already available for these loggers.

FIGURE 15-1. A light detector connected to a data logger will allow you to record the arrival of ocean waves like this one.

If a sensor produces an output voltage that doesn't exceed the allowable input voltage for the logger, no circuitry is needed (unless the signal is so small that it requires amplification). A typical logger has an allowable input range of 0 to 2.5 volts. This means you can safely and directly log the voltage of many kinds of disposable and rechargeable power cells and batteries.

Voltage Divider Sensor Interface

If your sensor is resistive it won't produce a voltage, so you'll need to connect it as half of a voltage divider. The simplest interface circuit for resistive sensors is the single resistor or potentiometer circuit shown in Figure 15-2. It can be used with light-sensitive photoresistors, temperature-sensitive thermistors, and other sensors having a resistance that changes with pressure, touch, weight, acceleration, rotary motion, and so forth.

R1 is a resistor or pot connected in series with the resistive sensor to form a voltage divider. Using a pot allows the sensitivity of the circuit to be easily adjusted. The free end of the sensor is connected to the logger's positive supply voltage, and the free end of R1 is connected to the logger's ground at the input. The junction of the sensor and R1 is connected to the logger's positive input. Be sure to check the polarity of the logger's input before making the connections.

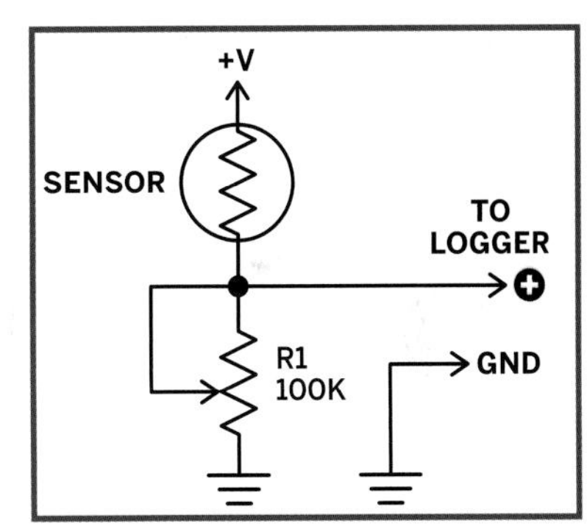

FIGURE 15-2. Use this simple voltage divider to connect resistive sensors, like photoresistors and thermistors, to voltage loggers..

It's best to use the logger's supply to power this interface, since this will avoid applying excessive voltage to the input. If the positive supply isn't available externally, you can open the battery hatch or the enclosure and carefully solder a connection wire to the positive (+) side of the battery connection.

Op-Amp Sensor Interfaces

If your sensor's output is too low for your logger, you'll need to amplify it. Operational amplifiers are ideal for boosting the very tiny current from light-sensitive photodiodes and the voltage produced by thermocouples in response to temperature. Few components are needed, since most of the electronics are inside the op-amp. For best results, select an op-amp that can be powered by a single polarity supply at or below the operating voltage of the data logger. This allows the sensor interface to be powered by the logger, and ensures that the sensor output won't exceed the logger's voltage.

You can get many different low-voltage op-amps from Maxim, Texas Instruments, National Semiconductor, and others. They're available in both traditional 7-pin mini-DIPs and surface-mount packages, from Jameco, Digi-Key, Mouser, and other distributors.

I've had very good results using the TI 271 with 7-bit Onset loggers. This op-amp requires a minimum of 4V, which exceeds the 3V lithium battery that powers these loggers, so I have to attach a 9V battery to the modified logger. I'm planning a new set of modified loggers that will use the TLC251, the TLC252 (a dual 251), or other op-amps powered by only a few volts.

Some sensors, like photodiodes and photovoltaic or solar cells, produce a variable current at a relatively stable voltage. Figure 15-3 is a circuit I've used to interface various photodiodes with 7-bit and 11-bit Onset Hobo loggers. The photodiode generates a small photocurrent when illuminated by a light source of the appropriate wavelength. The op-amp converts the current to voltage and amplifies it so it can be saved by a logger. The amplification or gain equals the resistance of feedback resistor R1. Thus, if R1 is rated 1,000,000 ohms (1MΩ), the photocurrent is amplified by 1 million. Reduce R1's resistance if the circuit's output approaches the power supply voltage before the light

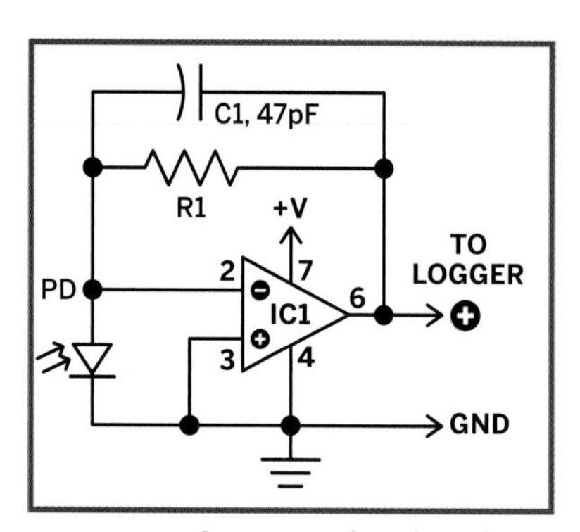

FIGURE 15-3. Current-to-voltage input for a data logger. PD is a photodiode. IC1 is a single, not dual, op-amp.

levels you're trying to record are reached. Increase R1 if the circuit isn't sufficiently sensitive. Capacitor C1 helps prevent oscillation of the circuit.

The circuit in Figure 15-4 is for sensors that produce a variable voltage instead of a current. Thermocouples, for example, produce a voltage proportional to the temperature to which they're heated. The gain of this circuit is 1 + (R2/R1). The values shown in Figure 15-4 provide a gain of about 1,000, suitable for monitoring the heat from a space heater when used with a type K thermocouple. Increase R2's resistance for more temperature sensitivity.

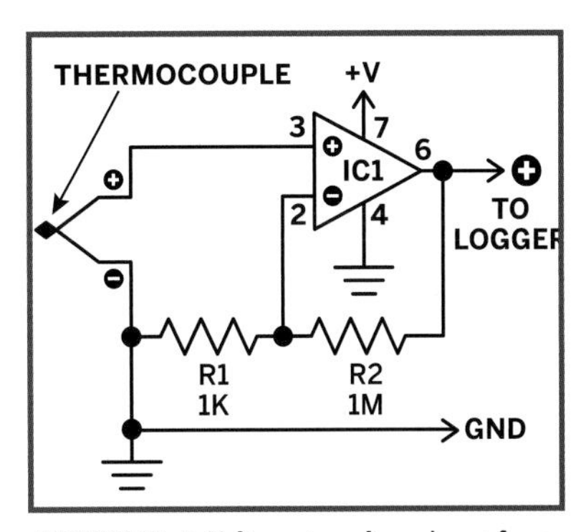

FIGURE 15-4. Voltage-to-voltage input for a data logger. IC1 is a single op-amp.

Some Data Logger Applications

I first began using standalone data loggers to measure the temperature of fire ant mounds, our doghouse, and the cells in a paper wasp nest. These experiments were so interesting that I decided to measure other parameters, especially the sun's ultraviolet and photosynthetically active radiation (PAR), the blue and red wavelengths that make plants grow. No data loggers were available for these applications, so I used the circuit in Figure 15-3 together with various DIY UV and PAR sensors.

FIGURE 15-5. Hobo data logger modified with UV detector (aluminum tube with Teflon diffuser cap) and current-to-voltage op-amp circuit on rectangular circuit board above the battery.

The most interesting results from these logging projects came during one of my annual trips to calibrate instruments at Hawaii's Mauna Loa Observatory (MLO).

Prior to the trip, I modified 15 Onset Hobo loggers with the op-amp circuit in Figure 15-3 and photodiodes fitted with UV-B filters (Figure 15-5).

I hid the modified loggers around the Big Island in places with full sunlight. Several days later, I retrieved the loggers, which provided a record of solar UV-B between sea level and the 11,200-foot elevation of MLO. The MLO logger showed that cumulus clouds near the sun caused UV-B increases of up to 15%. Figure 15-6 shows the data on a clear day and on a day with clouds during the afternoon. This finding led to a report in a leading scientific journal (F.M. Mims III and John E. Frederick, "Cumulus Clouds and UV-B." *Nature* 371, 1994).

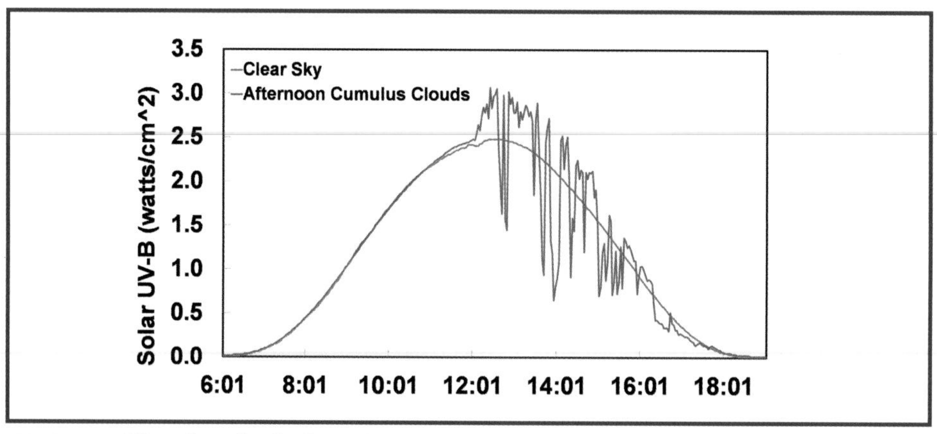

FIGURE 15-6. Effect of cumulus clouds on solar UV at Hawaii's Mauna Loa Observatory (11,200 feet).

In Hawaii I've also measured the sun's UV-B underwater and reflected by surf. Figure 15-7 shows a typical result when the sensor was mounted on an 11-foot pole and held over the surf: Significant UV-B is reflected from ocean surf. When I repeated these measurements over a turbulent waterfall in Colorado, very little UV-B was reflected.

During the May 10, 1994, annular eclipse of the sun, one of my modified Hobos monitored the sun's UV-B. During peak annularity, the sun formed a thin ring of brilliant light around the moon for about five minutes. The highly diminished solar UV during this time is indicated by the bottom of the dip in Figure 15-8.

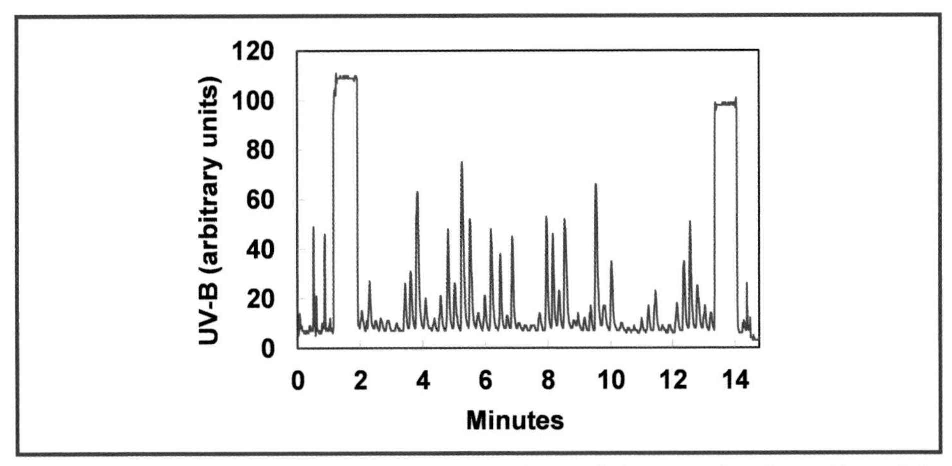

FIGURE 15-7. Solar UV reflected from waves breaking on lava in Hawaii. The rectangular spikes at either end of the chart show when the UV detector was pointed upward toward the sun and sky.

NASA twice sent me to Brazil to monitor the ozone layer and other atmospheric parameters during that country's annual burning seasons. Hobo data loggers modified to measure UV and PAR silently monitored whatever sunlight managed to leak through the smoky sky, allowing me to concentrate on measuring smoke and ozone. Figure 15-9 shows the PAR measured at Alta Floresta on a very smoky day and a cleaner day. When I left Alta Floresta for a remote camp on the Cristalino River, a concealed Hobo with a DIY sensor provided an important record of PAR during my absence.

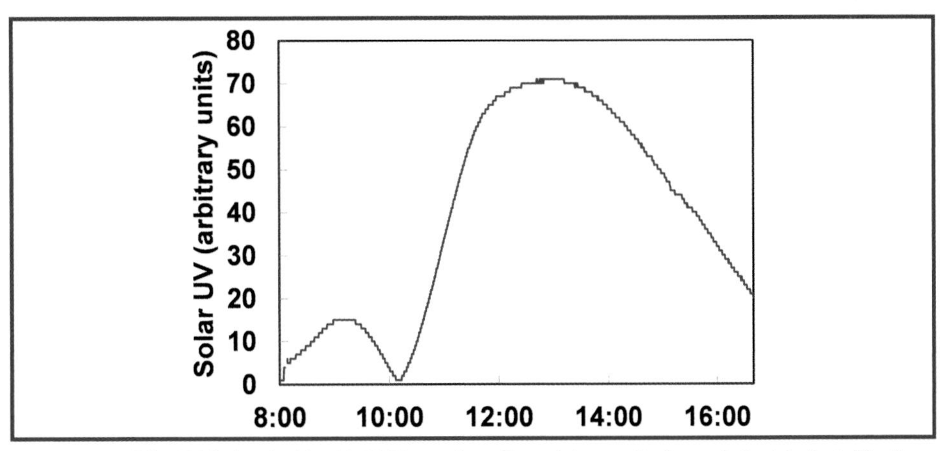

FIGURE 15-8. Solar UV during the May 10, 1994, annular eclipse of the sun. Peak annularity is indicated by the bottom of the sharp dip in the graph, when for about 5 minutes the only sunlight was a brilliant ring around the moon.

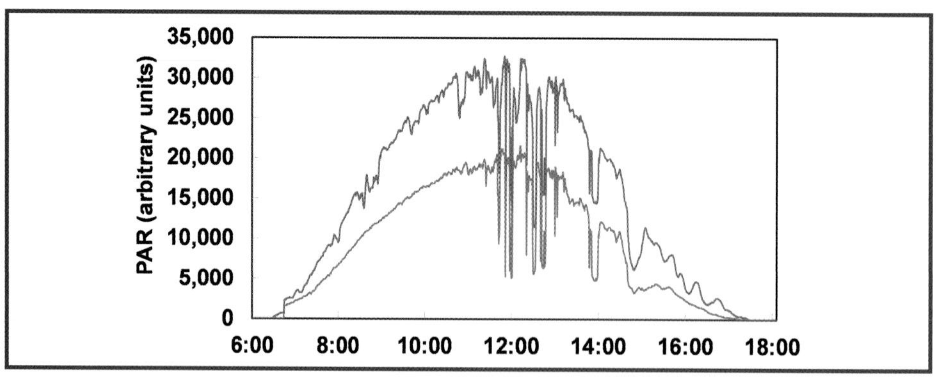

FIGURE 15-9. The photosynthetic radiation (PAR) responsible for plant growth on a very smoky day (blue) and a cleaner day (red) at Alta Floresta, Brazil. Sharp dips indicate clouds at the sun.

Figure 15-10 shows two pyranometers (solar radiation sensors) at Hawaii's Mauna Loa Observatory connected to an Onset U12-006. My recent work with these remarkable 11-bit loggers has led to new findings I hope to publish. The pyranometers were designed by my colleague Dr. David Brooks of the Institute for Earth Science Research and Education (IESRE) and are available as kits or assembled units at http://instesre.org/Aerosols/order_form.htm.

FIGURE 15-10. The sunlight intensity measured by two solar pyranometers at Hawaii's Mauna Loa Observatory is being logged by two of the four channels in an Onset 11-bit data logger.

Get Started Logging

If you want to do serious amateur science, data logging has huge potential. I recommend you acquire a basic temperature-sensing logger and start experimenting. The experience might inspire you to find entirely new logger applications.

How to Document What You Make or Discover

16

No matter what you plan to make, invent, or discover, one of the best steps you can take is to keep a detailed record of your progress. The traditional mechanism for documenting inventions and projects is a paper notebook, preferably one that is bound to assure that pages have not been added or removed. While paper notebooks remain very popular and are easy to update and store, the computer era has provided a variety of ways to record the progress of a project in far more detail and flexibility than possible with paper.

If you have commercial plans for what you're designing or making, a notebook will help establish your intellectual property rights and document your expenses. Be sure to enter all your ideas into your notebook, but keep in mind that ideas alone are not patentable. Your notebook should disclose how to transform an idea into a working apparatus, tool, rocket, instrument, robot, or computer program.

Keeping a Traditional Paper Notebook

During high school I kept notes about various projects on standard looseleaf notebook pages. Most of those pages were lost, including the ones that described an analog computer that could be programmed to translate 20 words of one language into another. When I donated this computer to the Smithsonian Institution in 1986, it was still in good condition, but there were no notes about its design and construction.

The lesson is clear: record details about your inventions and projects.

That's what I began doing while a senior at Texas A&M back in 1966. I recorded details about various experiments and projects in a sturdy, 108-page Aladdin spiral-bound notebook. Topics included miniature guided rockets, travel aids for the blind, lightwave communications, and how to use semiconductor diodes and solar cells to both emit and detect light (Figure 16-1).

William J. Holt was formerly a vice president for research at Varo, Inc., a manufacturer of infrared night-vision gear in Garland, Texas. When I showed Holt my spiral-bound notebook in 1966, he gave me a real laboratory notebook and some sound advice about how to document experiments and inventions. I still recall the essence of his helpful suggestions:

- Avoid using a spiral notebook for recording inventions and original research. Instead, use a bound notebook with gridded pages.
- Include plenty of detail about your experiment or project.

- Date each entry and sign each page of the notebook.
- If you make what might be an invention or discovery, ask at least one other person to sign that he or she has read and understood what you have described (Figure 16-2).
- Don't worry about neatness. It's best to simply enter everything you think is important, scribbles and all.

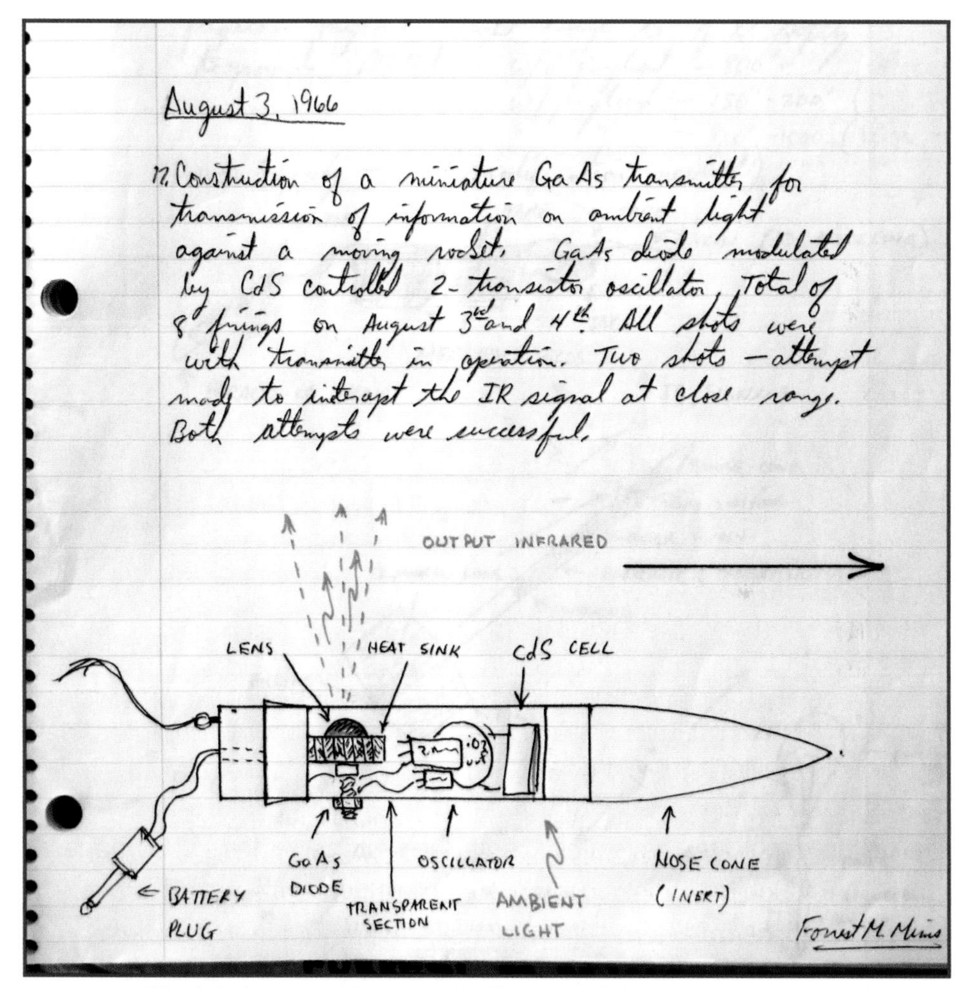

FIGURE 16-1. This relatively neat entry from my 1966 college notebook shows a rocket payload section that transmitted a signal to the ground over an invisible infrared beam from an early, high-power LED.

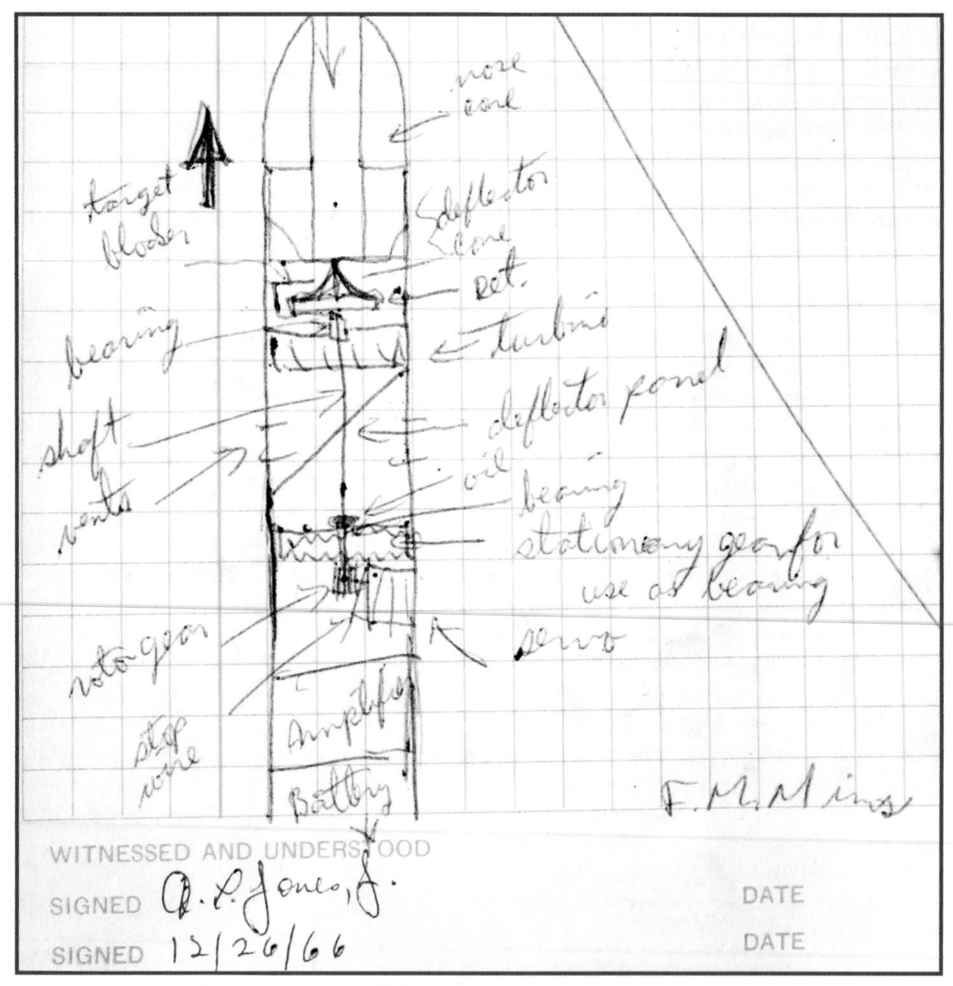

FIGURE 16-2. Notebook pages can also be "off-the-cuff" sloppy, like this sample from my first real lab notebook from 1966. More important than neatness is simply showing how to transform ideas into working projects, like this ram-air control mechanism to guide small rockets.

The lab notebook Holt gave me became the first in a series of 100-page, store-bought lab notebooks. From 1966 to 1996, I kept a record of many projects in these traditional lab notebooks. In addition to the projects described in the 1966 spiral notebook, topics included semiconductor lasers, remotely controlled cameras for making aerial photos from kites and balloons, and instruments that measured haze, water vapor, and the ozone layer. Also included were details about using LEDs and laser diodes as dual-purpose emitters and detectors of audio- and pulse-modulated light beams.

A completely unexpected spinoff from those lab notebooks occurred in 1979 when David Gunzel, then RadioShack's technical editor, came up with a new book idea. Gunzel suggested an idea for a hand-lettered book filled with electronic circuits that would resemble my lab notebooks. *Engineer's Notebook* sprang from Gunzel's idea and became the first of a series of 19 hand-lettered and illustrated books that sold millions of copies.

Keeping a Virtual Notebook

Paper notebooks retain important advantages, particularly when inventions are described on pages dated and signed by the inventor and witnesses. But when projects are best illustrated with photographs and charts, an electronic notebook provides far more flexibility.

Here are some tips for keeping an electronic notebook:

- Develop an organized virtual notebook containing separate folders for various projects, ideas, resources, expenses, and so forth.
- Save everything related to each project, including photos, videos, charts, data files, illustrations, circuit diagrams, and CAD files.
- Include plenty of detail about your experiment or project.
- Explore various ways for you and witnesses to digitally sign your entries.
- While you can create your own electronic notebook using standard office software, consider an online search to explore commercial or free electronic notebooks.

Going Further

Whether you keep a paper or electronic notebook—or both—be sure to carefully safeguard your notes by backing them up. Carbon paper may be old-fashioned, but it works well for backing up paper notebook pages. You can also photograph or scan paper notebook pages and save them electronically. That's what I'm doing with all my old notebooks (Figure 16-3).

FIGURE 16-3. The entire contents of my lab notebooks, thousands of pages of my atmospheric data (5 of 24 years shown at right), many thousands of scientific photographs, and much more, can be easily saved on just one flash drive (shown at right) or on a small fraction of the portable hard drive (center bottom).

As for electronic backups, I suggest storing your electronic notebooks on at least two external flash or hard drives. An online search will provide many more tips about keeping a laboratory notebook. Try searching "keeping a laboratory notebook." Our hardbound, gridded Maker's Notebook is a perfect paper option: www.makershed.com/products/makers-notebook-hard-bound.

17

MARS-BOT:

Adding Science to Robotics

Science fairs that involve engineering, physics, astronomy, and chemistry have declined in recent decades, while robotics competitions have rapidly grown in popularity. These contests teach students about electronics, mechanical engineering, and teamwork while providing plenty of fun in the process.

After watching a number of robotics competitions, I'm confident they can be expanded to include some science. So here's my proposal for a new kind of robot competition: Mars-bot, a simulated space mission.

Designing a Mars-bot Competition

A suite of sensors and samplers will transform a standard competition robot into a sophisticated Mars-bot that is guided across an imitation Martian landscape by a student mission team. Only the audience on the other side of the curtain will be able to see the Mars-bots—each team must rely on video images sent back by its Mars-bot to navigate across the Martian terrain, execute its mission, and return through the curtain.

Before the competition, each team receives an image of the Martian landscape, as if taken from a satellite, that identifies landmarks. A grid laid over the image provides dimensions, and the bot must visit and sample or measure the landmarks. If the competition is indoors, a bright overhead light can represent the sun. Scoring is based on the number of successful measurements and samples each Mars-bot returns, with ties broken by speed. Each team will create comprehensive mission reports and data analysis to receive academic credit.

FIGURE 17-1

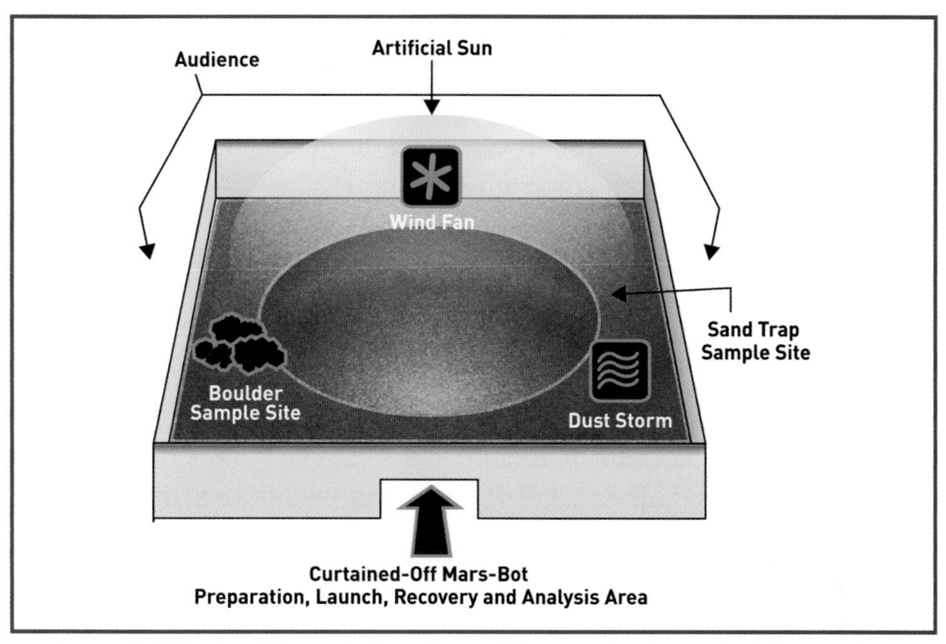

FIGURE 17-2. Each Mars-bot will be required to perform a minimum number of tasks after it disappears behind a dark curtain concealing the Martian landscape from the mission team.

Video Cameras

The most indispensable sensor on each Mars-bot will be at least one wireless, color webcam. The basic setup has a stationary mount that looks straight ahead. More sophisticated Mars-bots will feature a webcam that can rotate to better survey the landscape, find assigned goals, and provide visual clues when samples are collected. The camera itself can also report back data indicated by readouts or instruments in its field of view.

Simulated Dust Storm

Mars is known for its vast dust storms. A fan blowing dust across the path of a Mars-bot could test the ability of moving parts to survive a blast of grit. The reduction in electrical power that occurs when dust falls on a Mars-bot's solar panel can also be measured.

Wind-Speed Sensors

A Mars-bot should measure the speed of any wind it encounters. A fan or propeller mounted on the shaft of a small DC motor can act as a simple analog wind speed sensor. When wind rotates the motor's armature, a voltage proportional to the rotation rate will appear across the motor's terminals.

You can make a digital wind-speed sensor by mounting a disk on the shaft of a propeller. Glue a small magnet to the outer edge of the disk, and mount the assembly so the edge of the disk rotates past a Hall effect sensor. The Hall sensor will provide a voltage pulse each time the magnet passes by. If needed, you can mount additional magnets or weights around the disk to balance it. Calibrate the sensor by placing it adjacent to a commercial, handheld wind-speed sensor at various distances from a fan.

Haze

Dust blown high into the Martian atmosphere can cause long-lasting haze. In an indoor competition, periodic dimming of the artificial sun can simulate haze, while passing clouds will create the same effect during an outdoor competition. A photodiode or solar cell can detect the reduced light. Mount it behind a plastic diffuser to ensure it receives light no matter the artificial sun's location.

Temperature

The temperature during a mission will slightly change with wind, haze, and cloud conditions. You can easily measure it using a thermistor or integrated temperature sensor, or better yet, with an infrared thermometer, which can also scan the temperature of various objects along the mission course.

Spectrometer

The colors of rocks, sand, and soil provide important clues about their composition. The Mars-bot's video camera can be used as a simple 2-color spectrometer. Photo processing software can analyze individual video frames to express the relative intensity of the blue, green, and red wavelengths of the simulated Martian landscape.

Sand and Pebble Sampling

An especially important part of a Mars-bot mission is to collect geological samples and return them for analysis. Mission controllers will use their video link to steer their Mars-bot to sand and gravel sites along the course. The mechanical features of the current generation of competition robots can be easily modified for sand and pebble sampling.

Borer to Collect "Rock" Sample

The mission team will steer their Mars-bot to a mock boulder, bore a sample, and stash it for the return trip.

The boulder might be fashioned from a thin but rigid sheet of wood or other soft material. A standard battery-powered drill fitted with a ½" to 1" hole saw can be mounted on the front of the Mars-bot, and switched on and off by a radio-controlled relay connected across the drill's power switch. In my experience, a circle of wood removed by a hole saw stays inside the saw until it's manually removed, so the saw itself should hold and retain one or two thin samples.

These are just a few ideas; I would love to hear yours. I'm hopeful that science-oriented Mars-bots will significantly expand the hands-on educational experience already provided so well by established robotics competitions.

How to Photograph the Solar Aureole

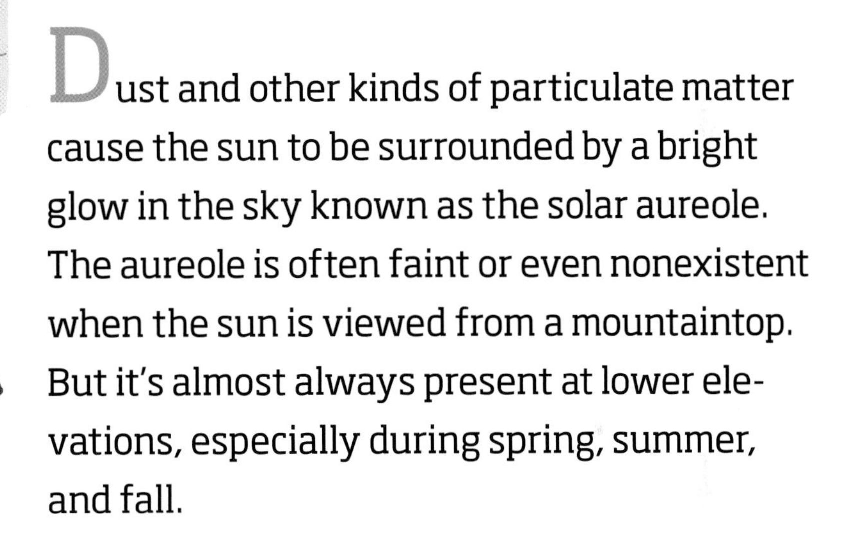

Dust and other kinds of particulate matter cause the sun to be surrounded by a bright glow in the sky known as the solar aureole. The aureole is often faint or even nonexistent when the sun is viewed from a mountaintop. But it's almost always present at lower elevations, especially during spring, summer, and fall.

The Solar Aureole

The diameter and brightness of the aureole is related to the scattering of sunlight caused by particulate matter. This means a record of solar aureole photographs can provide a good indication of the transmission of sunlight through the atmosphere. The color of the sky beyond the aureole also provides clues about stuff in the air.

Since 1990, I've made almost daily measurements of the ozone layer, solar ultraviolet radiation, haze, total water vapor, and other sun and sky measurements from a field adjacent to the small farmhouse that serves as my South Texas office. In 1998, I bought my first digital camera, a 1.5-megapixel Fuji MX-700. To date, this camera has provided 7,619 images (1,280×1,024 pixels) of the solar aureole and the sky over the north horizon. While the resolution is low by today's standards, it's more than adequate for a record of sky images.

These solar aureole images provide important information about my electronic sun and sky measurements, for they quickly reveal the presence of thin clouds or haze that might have affected the measurements. They also provide a visually convenient way to compare the clarity of the sky across the seasons and years.

Photographing the Aureole

Solar aureole photos can be made with virtually any kind of digital camera. For serious studies, you'll want to use a camera that allows you to set the same exposure duration and f-stop for all your aureole photos. This will provide a record of the sky without unwanted automatic adjustments by the camera that alter or even remove the changes in sky brightness and color caused by dust, smoke, and other forms of air pollution.

The aureole is washed out if the sun is photographed directly. Worse, your eyes and a digital camera's image sensor can be permanently damaged by the focused image of the sun. Therefore, it's necessary to design an occulting device that blocks the direct sun when making solar aureole photos. To protect your eyes, it's also necessary to use a method that does not require you to look anywhere near the sun.

I've developed various methods for photographing the solar aureole in which you need only look at your camera. My favorite occluder rig is a simple camera platform that keeps the sun and occluder in the same position for each photograph. This greatly simplifies the comparison of photos.

You can design your own platform or you can try my simple version, the Solar Photography Occluder Rig, explained in the next section, "DIY Solar Photography Occluder Rig."

Making a solar aureole photo with the occluder in place is simple. If the sky is not overcast, put on sunglasses and a hat and go outdoors with your camera mounted on the occluder. If it's awkward to sit in a chair or on the ground, brace the occluder rig against a stable object such as a fence or wall.

Switch on the camera, point it toward the sky away from the sun, and adjust its position so the occluder is centered in the display. Then look at the front of the camera—not at the sun—while pointing the camera toward the sun. When the shadow of the occluder ball falls directly over the lens, press the shutter button and quickly move the camera away from the sun.

DIY Solar Photography Occluder Rig

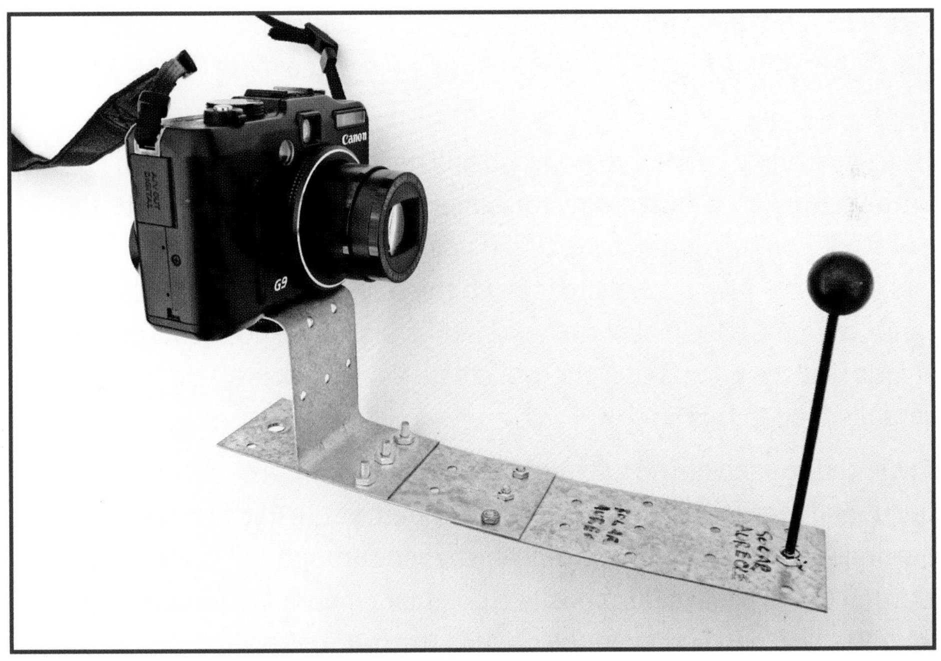

FIGURE 18-1. A Canon G9 mounted on a simple occluder platform that blocks direct sunlight from entering the camera lens.

Materials

- Three 1¾" × 5" steel mending plates or equivalent
- Six #6-32 screws
- Eight #6-32 nuts
- ¼-20 bolt
- 1" threaded rod
- Wood ball slightly larger than the diameter of the camera's lens (found at craft or hobby stores, wood balls may come with or without a hole bored partway or entirely through)
- Black ink or paint (a black marker pen works fine)
- Drill and drill bits

Join two of the mending plates end-to-end with three 6-32 screws and nuts, as shown in Figure 18-1. (You may need to slightly enlarge the holes in the mending plates.) Drill out the center hole near the end of the third plate to ¼" diameter, to accept the ¼-20 bolt that will secure the camera. Bend the third plate twice as shown above and fasten it to the center of one of the two base plates with three more 6-32 screws and nuts.

Next, determine how high the wood ball needs to be so that it's in line with the camera lens. Mount your camera atop the bent plate with the ¼-20 bolt. Hand-tighten the bolt to avoid damaging the camera; if it's loose, insert a few washers to take up the slack. Now measure the distance from the base plates to the center of the lens. Add to this distance 1" plus half the diameter of the wood ball. Use a hacksaw to cut the threaded rod to this length.

Twist the cut end of the rod into the wood ball. Turn a 6-32 nut onto the uncut end of the rod until it's about 1" from the end. Remove the camera and set it aside. Insert the end of the rod through the center hole in the end of the base plate opposite the camera mount and secure it with another 6-32 nut. Complete the occluder by coating the ball and the threaded rod with black paint or ink.

Never look at the sun while making aureole photographs. To avoid sunlight damage to your camera's image sensor, you must work fast. You may void the camera's warranty if you damage it by pointing it at the sun.

Doing Science with Your Images

For serious scientific purposes, it's best to make solar aureole photos at the same manual settings (I use 1/1,600 at f/4) and at the same time each day the sun is visible.

Since 1998, I've made 5,764 aureole photos at or near local solar noon. Solar noon varies during the year, in accordance with the equation of time. You can find solar noon calculators and tables for your location online. Just search Google for "solar noon." Some sundial sites also have solar noon tables.

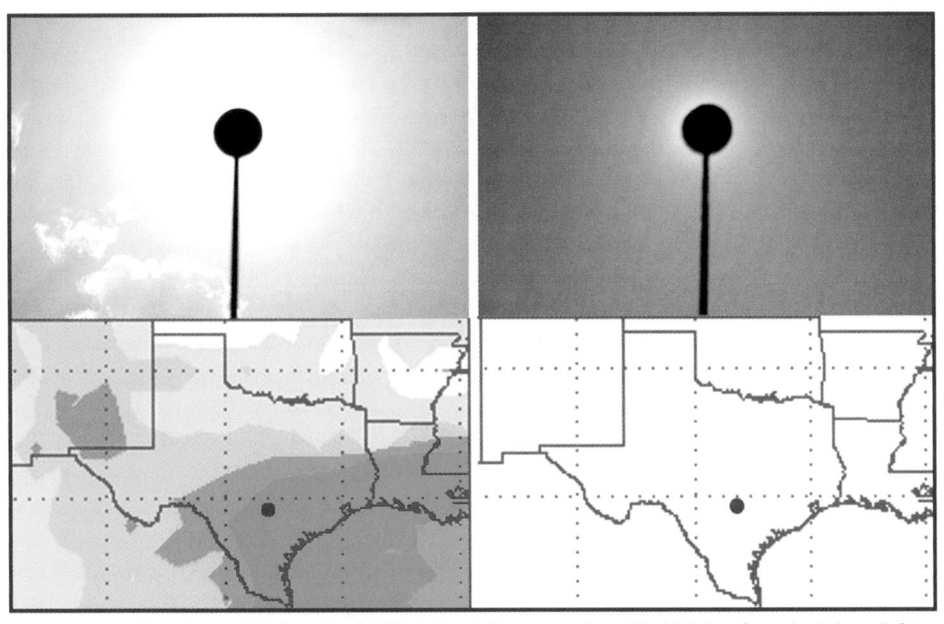

FIGURE 18-2. The solar aureole photographed from South Texas on a day with thick dust from the Sahara (left) and a very clear day (right). Both images confirm the Naval Research Lab's aerosol forecasts for the same days (http://nrlmry.navy.mil/aerosol/).

An even better choice would be to photograph the aureole when the sun is at the same angle in the sky. Again, the web has various sites that provide solar angle calculators. A good choice would be to select the sun angle at noon on the winter solstice when the sun reaches its lowest point in the sky.

Going Further

In the next chapter, we'll use a free image-processing program to analyze your solar aureole images and other kinds of photographs.

Record Your World from a Picture Post

19

Scenic photos of storms, clouds, sunsets, mountains, and beaches are loaded with data about the natural world. Even pictures of backyards and parks can provide valuable environmental information.

For decades, NOAA and NASA satellites have collected images of the Earth from space. In recent years, ground-based webcams have captured views of scenic sites around the world. But organized sequences of high-resolution landscape and sky photos taken of the same scene over several years or more are not that common. This lack provides an important opportunity for amateur scientists to fill a huge data void. All that's necessary is a digital camera and a platform to place it on, so that it can be used to collect images at regular intervals.

The Picture Post Project

John Pickle, whose background includes geology and meteorology, is on the science faculty of Concord Academy in Concord, Massachusetts. In 2005, Pickle began Picture Post, a project carefully designed to collect periodic sets of landscape images using a digital camera pointed in eight different angles from an octagonal platform mounted atop a post.

The Picture Post project made total sense to me. Since the fall of 2000, I've taken almost daily digital fisheye photographs from a post in a 1.5-acre field adjacent to my office. So far I've collected more than 7,842 images of the sky above and the horizon around the post.

FIGURE 19-1. The north horizon at the Mims place photographed just after solar noon on the same February day in 2002 (left) and 2011 (right). Note the growth of the two live oak trees during the nine-year interval. The contrails in the 2011 image were a result of a damp sky.

While these photos were originally intended to record haze and clouds in the sky, it soon became obvious the images were also recording changes in the trees and other vegetation around my site as well as a nearby field and farm. Seasonal changes were very obvious, especially the transformation of the landscape each spring and fall. Over the years, changes in the height of trees were noticeable. Those surface changes were accompanied by sky changes, especially haze from Asian dust and Mexican smoke during spring and African dust during summer and fall.

FIGURE 19-2. This sequence of images from my instrument and camera post shows the growth, bloom, and decline of a patch of native sunflowers. The central trail was kept open to record the north horizon.

I didn't fully appreciate the environmental record these images provided until learning more about Picture Post, which is part of Digital Earth Watch (DEW), a NASA-sponsored partnership of various organizations. According to the project's website, "Picture Post was created for DEW as a tool for nonscientists to monitor their environment and share their observations and discoveries."

How to Make and Install a Picture Post

My original *picture post* is just a 6' length of 4×4 lumber topped by a 17" length of 2×6 lumber that serves as an instrument and camera platform. Both are pressure-treated to prevent rot. Pickle's Picture Post design is better suited for recording both the surrounding landscape and the overhead sky.

While Pickle also uses a 4 × 4 post, he recommends a length of 7' to 8', 4' of which will be above ground. His camera platform is much more than a flat surface—it incorporates a raised octagon designed to permit a camera to capture a 360° sequence of images around the post. Hold the camera with its back against each segment of the octagon, and make an exposure. Take a ninth photo with the camera facing straight up.

Although you can use ordinary pressure-treated lumber for the picture post, Pickle recommends a plastic composite post because of its virtually unlimited life. The top of my post is chest high (about 4', the same height recommended by Pickle), although young children might require a shorter post. You could use a step stool, but this may not be practical for all outdoor situations.

You can make a camera platform for the post from two pieces of weather-resistant wood or plastic. Pickle recommends making the platform base by cutting a 9" circle or octagon from ¾" plywood. Cut the camera octagon from 2" or ½" plywood with 2" sides, measuring 5" across facing sides. Center it over the 9" base, and screw or glue it to the base with water-resistant adhesive. If you make the platform from wood, Pickle recommends waterproofing it with several coats of polyurethane. For full details, see picturepost.unh.edu.

To save time, you can buy a ready-made plastic camera platform (with a raised octagon and a recessed *N* to indicate north) for $25 from the Picture Post website.

When the camera platform is ready, it's time to install the post. Select a site that provides the best view of the surrounding landscape consistent with the vertical view. While you might be able to mount the camera platform on a fence post, deck railing, or other outdoor support, installing your own post provides more flexibility. A location that features views of both a distant horizon and nearby plants will provide a diverse record of the landscape. A vertical view that includes a bit of tree canopy will let you record seasonal changes while also tracking cloud cover.

Dig a hole at least 3' deep, and wide enough for the 4 × 4 post. If you live in an area where the soil freezes, the hole should be deeper than the frost level. For example, in northern Minnesota, Pickle recommends the bottom of the post be 5' below the surface (in which case, to maintain 4' of post above ground, you'd need an overall length of 9').

Place the post in the hole and roughly align it so that each side faces a cardinal direction (north, south, east, and west). Backfill the hole with soil while using a level to keep the post as vertical as possible. While the hole is being filled, tamp down the soil with your feet until it's firm and the post is stable.

The camera platform should be centered on top of the post and attached with four 3" to 3½" coarse-thread exterior drywall screws.

Four matching pilot holes should be drilled into the top of the post, but only after it's oriented so that one side of the octagon faces due north. You can use a compass to find magnetic north, but you'll need to correct the reading to find true north.

Pickle recommends using the declination chart at www.thecompassstore.com/decvar.html.

How to Use a Picture Post

The Picture Post program doesn't require daily photographs, but their website recommends that you take a set of photos every week or two when possible. Based on my results, it seems best to take photos at solar noon (when the sun is at its highest point in the sky). If that's not feasible, take all photos at about the same time of day.

FIGURE 19-3. Concord Academy student Will Jacobs takes a Picture Post photograph.

In my experience, more frequent photographs are important for capturing days with clear skies, haze, and cloud cover, and the rapid changes that occur to deciduous vegetation during spring and fall. Here are just a few examples of what you can do:

Expand your image collection by taking more frequent photos during spring and fall.

FIGURE 19-4. Seasonal changes looking northwest from the Concord Academy Picture Post. Measure plant growth by taking daily images during spring. Take frequent sky photos, since the sky changes much more rapidly than the landscape.

You can supplement your Picture Post archive by posting your photos on a website. Add information about data and observations from your site.

Automated webcams provide a convenient way to record changing landscapes and sky. An archive of webcam images can be made into a movie that will show these changes over time.

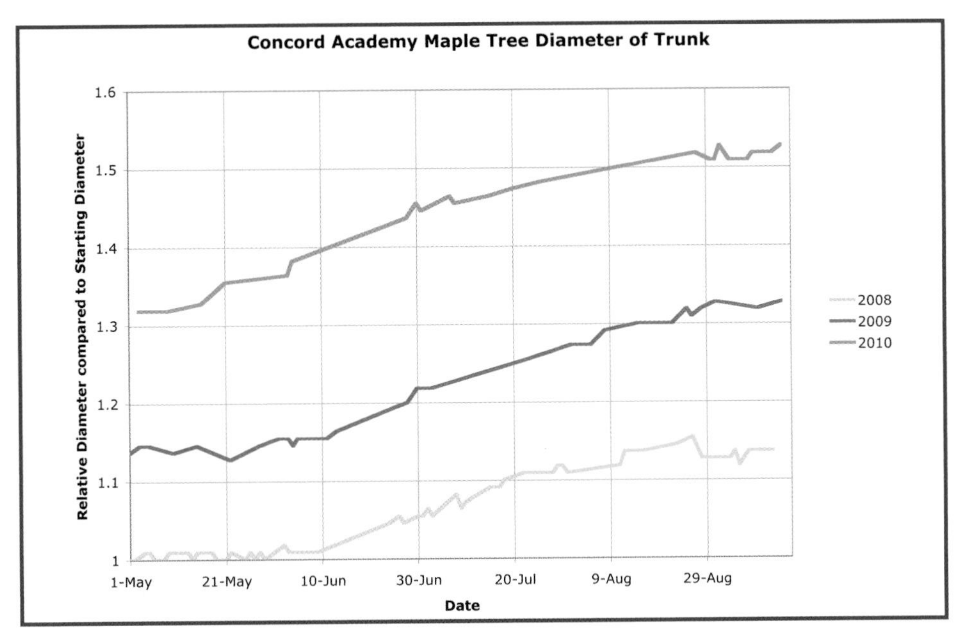

FIGURE 19-5. Growth of a maple tree extracted from three years of Picture Post images by John Pickle.

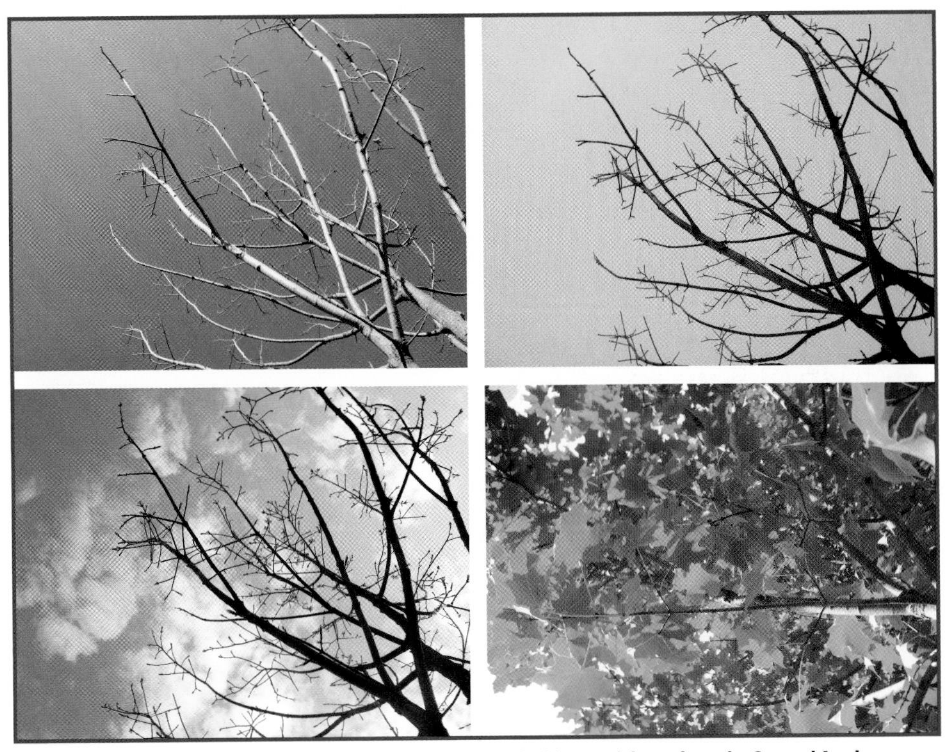

FIGURE 19-6. Seasonal changes of the sky and a nearby tree looking straight up from the Concord Academy Picture Post.

How to Participate in the Picture Post Program

Anyone can participate in the Picture Post program. You can register your picture post on the Picture Post website (http://picturepost.unh.edu/index.jsp) and even submit your photographs.

This will place your images into an archive for study by both amateur and professional scientists, and anyone else.

Digital Pinhole Photography

20

Long before the digital photography era, I enjoyed making photographs with a 35mm film camera equipped with a pinhole instead of a lens. The editors of *Popular Photography* magazine liked the results and published my article "The Pinhole: A 'Lens' That Just Won't Quit."

Much has happened since that article was published back in April 1974. A renaissance of sorts has occurred in pinhole photography, and Justin Quinnell is among its leaders. His website http://pinholephotography.org is loaded with hints, tips, and unique pinhole images. (For an inspirational video about Justin's work, see https://thelifeofdocumentary.wordpress.com/2012/06/29/the-life-of-a-pinhole-photographer/.)

Suitable Cameras

Any camera with a removable lens has potential for pinhole photography. Conventional film cameras can be used, but digital cameras with removable lenses are ideal. The exposure time can be easily changed, and results are available instantly. A new entry-level digital SLR costs $500 or more, but you might find a used one for considerably less.

Characteristics of Pinhole Images

Pinhole images are fuzzier than those made through a lens, but that can be an asset. If the fuzz is excessive, you can sharpen your images by using a smaller pinhole or photo processing software. Figure 20-1 shows the sharpening that resulted from reducing the pinhole size from 0.6mm to 0.3mm.

Another characteristic of pinhole cameras is nearly infinite depth of field. Make a pinhole photo of a very close object with a distant building or mountain in the background—it's all in focus. You can even use the sun for the distant object (Figure 20-2), but it will be fuzzy unless the exposure is brief.

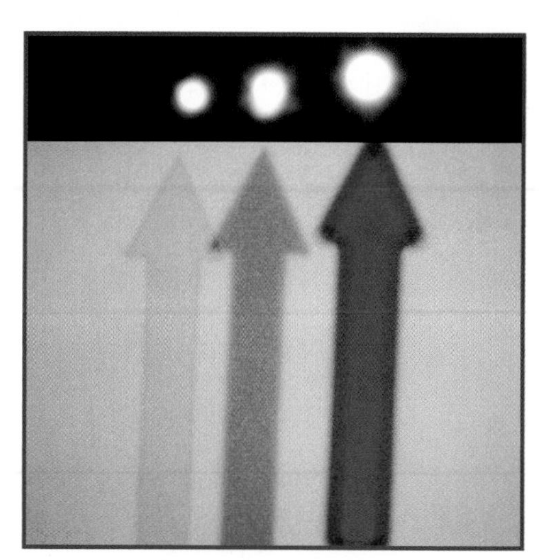

FIGURE 20-1. Three images of the sun and an arrow on a computer screen, photographed through three pinholes mounted on a Canon 40D digital camera. The largest pinhole (the width of a 0.6mm-wide pin) produced the brightest but fuzziest images (at right). The smallest pinhole (0.3mm) produced the dimmest but sharpest images. Note: All images except the figure on the right were enhanced using Microsoft Digital Image Pro software.

Pinholes punched through foil or thin metal leave behind a projection of torn metal on the exit side known as a *crown burr*. Pinhole photographers often remove the burr with sandpaper. When left in place, the burr can cause uniquely beautiful effects, especially when making pinhole photos of the sun, as shown in Figure 20-3.

FIGURE 20-2. Handheld beverage-can pinhole image of barbed wire illuminated by flash and the morning sun (1/60 sec., ISO 320).

FIGURE 20-3. Sun Spray, a handheld image enhanced only by the crown burr formed in the aluminum foil pinhole (1/8 sec., ISO 100).

Pinhole Exposure Times

Pinholes admit much less light than a conventional camera lens, so exposures must be longer. This usually means the camera must be mounted on a tripod or placed on a stable surface. But thanks to the high sensitivity of digital cameras, handheld photos are often possible when the scene is brightly illuminated. I've made handheld pinhole photos at speeds from 1/30 second (bright sunlight) to 1/8,000 second (the sun itself).

Pinhole Tips & Samples

The best advice for the new pinhole photographer is one word: experiment. Try various pinhole sizes, mounting methods, and distances from your camera's sensor. Be careful to keep dust off your camera's sensor when removing the lens to install your pinhole.

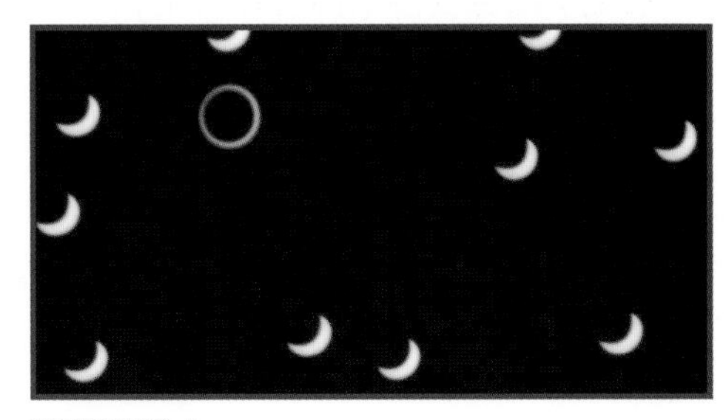

FIGURE 20-4

As for subjects, the world is your limit. Unlike a camera with a lens, it's easy to include the sun in pinhole photos. Just remember never to look directly at the sun through a pinhole camera.

FIGURE 20-5. Three dramatically different views of a high-voltage power transmission tower, all made without a tripod.

How to Make a Foil Pinhole

The pinhole images I published in *Popular Photography* and my recent eclipse images (Figure 20-4) were all made by pushing an ordinary 0.6mm-diameter pin entirely or partially through aluminum foil. Here's how:

1. Use scissors to cut a square of aluminum foil large enough to cover the lens opening of your camera. Heavy-duty foil is best, but standard foil is OK.

FIGURE 20-6

2. Place the foil on a desk protector, place mat, or other flat substrate that has a slightly resilient surface. Smooth the foil by rubbing it with the tip of your finger.

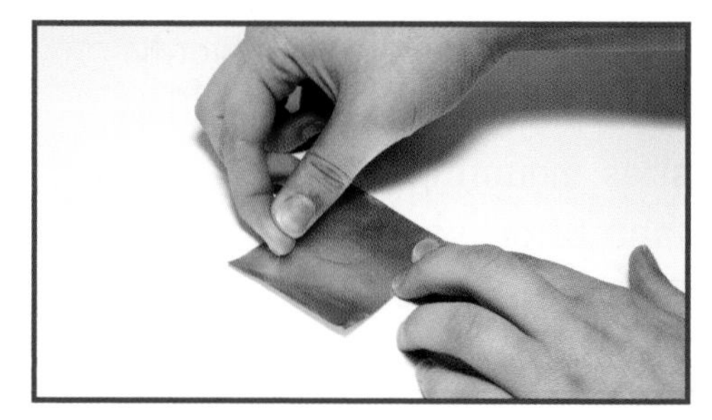

FIGURE 20-7

3. Carefully press the pin into but not completely through

the foil. For initial experiments, the diameter of the hole should be about half the diameter of the pin.

4. Remove the lens from your camera and place the foil over the lens opening with the pinhole roughly centered. The topside of the foil should face away from the camera. Use masking tape to secure the foil in place. Be sure no stray light can enter the camera; it will wash out your images.

FIGURE 20-8

FIGURE 20-9

How to Make a Better Pinhole

For optically cleaner images, a pinhole formed in thin sheet metal is best. This method is easiest to implement by mounting the metal onto a camera body cap that is placed over the lens opening when the lens is removed. Aluminum or copper sheet metal from a hobby store will work, but the simplest and cheapest source is an aluminum beverage can, pie tin, or food tray.

Pinhole photographers use various methods to form pinholes in sheet metal (search Google for details). I prefer the "brute force" approach, as detailed here

1. Rinse out an empty beverage can with water. Carefully cut the top and bottom off, cut lengthwise, and flatten to create one metal sheet.
2. Use old scissors to cut several 1" squares from the flattened can.

FIGURE 20-10

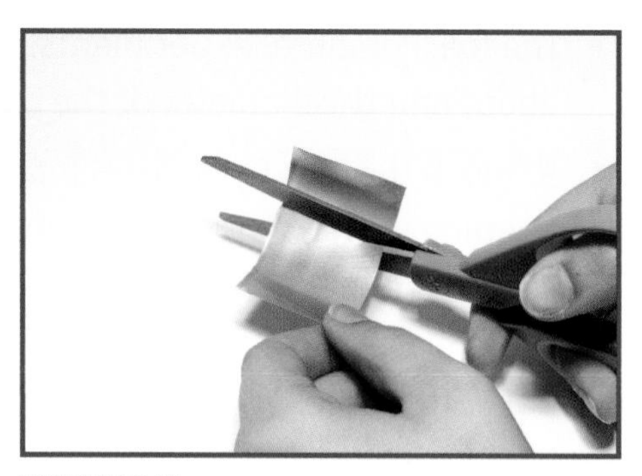

FIGURE 20-11

3. Place a metal square on a flat wood surface. Put the point of a straight pin at the center of the square and hold it in place with pliers. Lightly strike the head of the pin with a small hammer so that the pin just pierces the metal to form a circular hole about half the diameter of the pin.

FIGURE 20-12

4. Place a sheet of 220-grit sandpaper on a flat surface, business side up. Rub the backside of the metal square against the sandpaper to remove the crown burr, using circular strokes.

FIGURE 20-13

5. Look through the pinhole to check its uniformity. About two-thirds of my pinholes made in this fashion appear perfectly circular, which is what you want.
6. Bore a ¼" hole in the center of the camera's body cap.

FIGURE 20-14

FIGURE 20-15

7. Place a metal square with a pinhole over or behind the body cap so the pinhole is centered in the ¼" aperture. Secure it tightly with removable masking tape so you can try other pinholes later. When you find the best one, secure it with adhesive.

FIGURE 20-16

Experiment!

Explore the tutorials at http://pinholephotography.org and other pinhole photography websites. After you learn the basics, mount your pinhole camera on a rigid tripod and try making portraits of perfectly still friends and relatives. Make a time exposure of the movement of the stars across the night sky. Or mount a pinhole on a light-tight extension tube to make a telephoto pinhole. Once you start making pinhole images, you'll soon think of many other ideas.

How to Make and Use Retroreflectors

21

A retroreflector is an optical device that returns an oncoming beam of light back to its source. It can be as simple as a tiny glass sphere or a type of prism formed from glass or plastic.

While ordinary flat mirrors also reflect light, the light isn't reflected back to the source but off to the side at the same angle the beam arrived. Only if the light beam is perfectly perpendicular to the surface of a flat mirror does it act like a retroreflector.

Retroreflectors are so much a part of everyday life that typically they don't attract much attention. But they attract plenty of attention while driving at night, when they seem to be almost everywhere. They're incorporated into the taillights of vehicles, safety barriers, traffic signs, and the painted stripes that separate lanes of traffic. Recently while waiting at a traffic light on a dark night, I noticed seven brightly glowing traffic signs coated with retroreflective paint. These signs were illuminated by the headlights of my pickup, and they each reflected the oncoming light back toward me.

A nearly full moon was overhead, and it has retroreflective properties, too. That's because the Apollo 11, 14, and 15 astronauts left arrays of precision retroreflectors on the moon. For 40 years, various observatories have pointed powerful laser beams at the moon and detected the light returned by those retroreflectors to accurately measure the distance between Earth and the moon.

Natural Retroreflectors

Retroreflection was observed long before artificial retroreflectors were invented. That's because the eyes of many animals double as retroreflectors that glow at night when viewed from the same direction as a fire or lamp. When I was a Boy Scout I observed the eyes of alligators in a Florida lake, glowing bright orange in the light of a campfire. Years later I used a flashlight to find caimans in Brazil's remote Cristalino River. Drivers notice the eyes of animals at night glowing brightly in the headlights of their cars.

The retroreflection exhibited by animal eyes is called eyeshine. A bright, head-mounted light provides the best way to observe eyeshine. Retroreflection from an eye occurs when some of the light focused onto the retina by the lens is reflected back through the lens, where it is refocused into a narrow beam that travels back to the source of light. While this occurs whenever the eye is opened, we notice it only at night when a bright light source held close to our eyes is pointed at the eyes of an animal. The retina alone is not a particularly good reflector, but in many vertebrate animals it is backed by a highly reflective layer of tissue called the tapetum lucidum.

The red glowing eyes of people in flash photographs is known as red eye. Human eyes lack a reflective tapetum, so human eyeshine is not nearly as bright as that of animals. The best way to eliminate red eye is to move the flash away from the camera's lens. The light will be reflected back toward the flash, and most of it will miss the lens.

Manufactured Retroreflectors

For decades highway signs and painted stripes on roadways have been coated with tiny retroreflectors. The earliest and still the most common are clear glass beads, which can be sprayed or poured over freshly painted road stripes and highway signs. Various kinds of reflective sheets are also used to coat warning barriers and signs. Some employ a layer of glass beads, while others use sheets embedded with tiny plastic corner prisms called microprisms.

Microprisms are tiny versions of much larger retroreflectors cut from a corner of a solid cube of glass or silica. These reflectors are called corner cubes or cube corners, and they provide the best performance. The retroreflector arrays on the moon are silica corner cubes. Retroreflectors can also be made by mounting three mirrors to form an open corner of a cube.

How to Make Retroreflectors with Glass Beads

Retroreflective glass beads, like those used on highway stripes, are available from various sources. I bought an 8oz bag of standard glass beads for $6 (plus shipping) from http://colesafety.com/.

Any surface that can be painted can be made retroreflective by sprinkling glass beads onto freshly applied paint. Here's how I transformed a plain wood letter "M" from a hobby shop into an attention-attracting object.

1. Place the wood letter on a sheet of paper and apply a thick coat of white enamel.

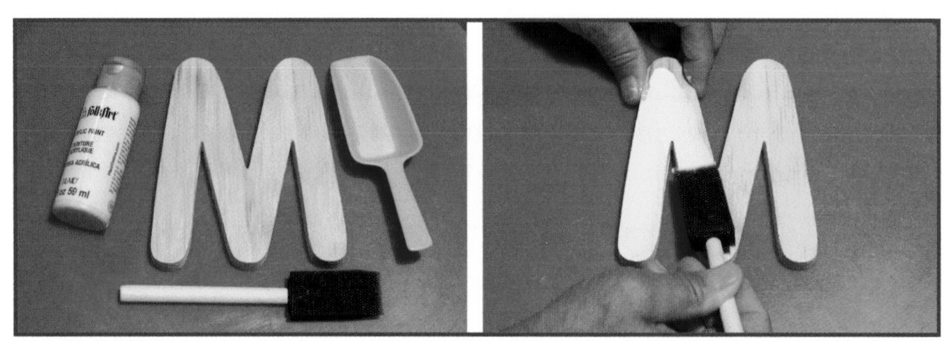

FIGURE 21-1

2. Pour glass beads over the entire painted surface of the letter. Be generous to make sure the entire painted surface is covered.

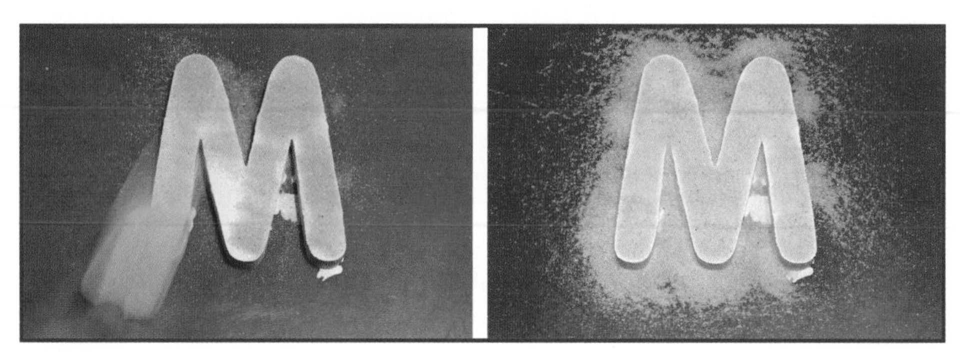

FIGURE 21-2

3. After the paint is totally dry, lift the letter from the paper and gently brush away the loose beads with a finger.
4. Pick up the corners of the paper and pour the unused beads into a container.

5. Place the completed letter in a dark room and shine a flashlight on it from 10' or more away. When the light is held near your face, the bead-coated letter will glow bright white.

FIGURE 21-3

How to Make an Open Corner Cube Retroreflector

A simple retroreflector can be made from three mirrors arranged to form the corner of a cube. You'll need three 2"×2" glass mirrors, double-sided tape, and a standard 2⅛"×2⅛"×4⅛" plastic box, all available from a hobby shop. Use care working with small mirrors, as they have sharp edges and are easily broken; supervise children. Follow these steps:

FIGURE 21-4. Look inside the reflective corner of the cube. With one eye closed, your open eye will be directly centered at the apex of the corner. It will stay there even when you move the cube at various angles.

- Clean your hands and work surface. Hold the mirrors by their edges and use glass cleaner to clean their surfaces.

- Place a 1" strip of double-sided adhesive tape on the back of a mirror and insert the mirror, shiny side up, inside the box lid.
- Place double-sided tape along the lower half of the back of a second mirror and stick it against one of the inner sides of the lid.
- Stick the third mirror adjacent to the second one so that all three mirrors merge to form a corner inside the lid.

Use suitable laser goggles and avoid eye damage by never looking directly at a laser beam or its reflection.

FIGURE 21-5. If the sides of the plastic container lid are slightly angled, the reflection from the cube corner will form separate beams that can be seen as three spots of light on a white wall behind the light source. When the sides of the box are perfectly square, these separate spots will merge into a single beam.

Going Further

Retroreflectors are ideal for aligning laser intrusion alarms and communication systems.

Plastic microprisms are excellent retroreflectors, but they require special equipment to make—until now. It should be possible to make arrays of microprisms using 3D printing with a transparent resin. Since such arrays wouldn't be restricted to flat sheets, as with mass-produced microprism arrays, there could be some intriguing applications for 3D arrays of microprisms.

Studies of eyeshine could make interesting science fair projects. For example, populations of spiders and moths that exhibit eyeshine can be surveyed with the help of a head-mounted light on a dark night. Spiders depend on insects for food, so their population is closely related to availability of insects. During a recent major drought where I live, the population of many insects plunged to nearly zero—and so did the spider count. Where there were once dozens of brightly glowing spider eyes each night, only one or two remained during the drought.

How to Use LEDs to Detect Light

22

Since an electromagnetic telephone receiver can double as a microphone, can a semiconductor light detector double as a light emitter?

That question was on my mind when I was a high school senior in 1962. Back then I didn't know that quantum effects in a semiconductor are unrelated to the electromagnetic operation of a telephone receiver. If I'd known that, I never would have connected a spark coil across the leads of a cadmium sulfide photoresistor to see if it would emit light. It did—a soft green glow punctuated with bright flashes of green.

During college I found that a silicon solar cell connected to a transistor pulse generator emitted flashes of invisible infrared that could be detected by a second solar cell. In 1972, I used near-IR LEDs and laser diodes to send and receive voice signals, through the air and through optical fibers. Later I experimented with two-way optocouplers made by taping a pair of LEDs together so they faced one another.

In 1988, I tried LEDs as sunlight detectors. They worked so well that the first homemade LED sun photometer I began using on Feb. 5, 1990, is still in use today.

Why Use LEDs as Sensors?

Silicon photodiodes are widely available and inexpensive. So why use LEDs as light sensors?

LEDs detect a narrow band of wavelengths, which is why I call them spectrally selective photodiodes. A silicon photodiode has a very broad spectral response, about 400nm (violet) to 1,000nm (invisible near-IR), and requires an expensive filter for detecting a specific wavelength.

The sensitivity of most LEDs is very stable over time. So are silicon photodiodes—but filters have limited life.

LEDs can both emit and detect light. This means an optical data link can be established with only a single LED at each end, since separate transmitting and receiving LEDs aren't needed.

LEDs are even more inexpensive and widely available than photodiodes.

Drawbacks of LEDs as Light Sensors

No sensor is perfect.

LEDs are not as sensitive to light as most silicon photodiodes.

LEDs are sensitive to temperature. This can pose a problem for outdoor sensors. One solution is to mount a temperature sensor close to the LED so a correction signal can be applied in real time or when the data are processed.

Some LEDs I've tested do gradually lose their sensitivity.

LEDs Detect Specific Colors of Light

The typical human eye responds to light with wavelengths from around 400nm (violet) to about 700nm (red). LEDs detect a much narrower band of light, having a peak sensitivity at a wavelength slightly shorter than the peak wavelength they emit. For example, an LED with a peak emission in the red at 660nm responds best to orange light at 610nm.

The spectral width of light emitted by typical blue, green, and red LEDs ranges from about 10nm—25nm. Near-IR LEDs have a spectral width of 100nm or more. The sensitivity of most LEDs I've tested provides ample overlap to detect light from an identical LED.

Figure 22-1 shows the spectral response of seven blue, green, red, and near-infrared LEDs that replace the usual silicon photodiodes and filters in my modified Multi-Filter Rotating Shadowband Radiometer, used for solar spectroscopy.

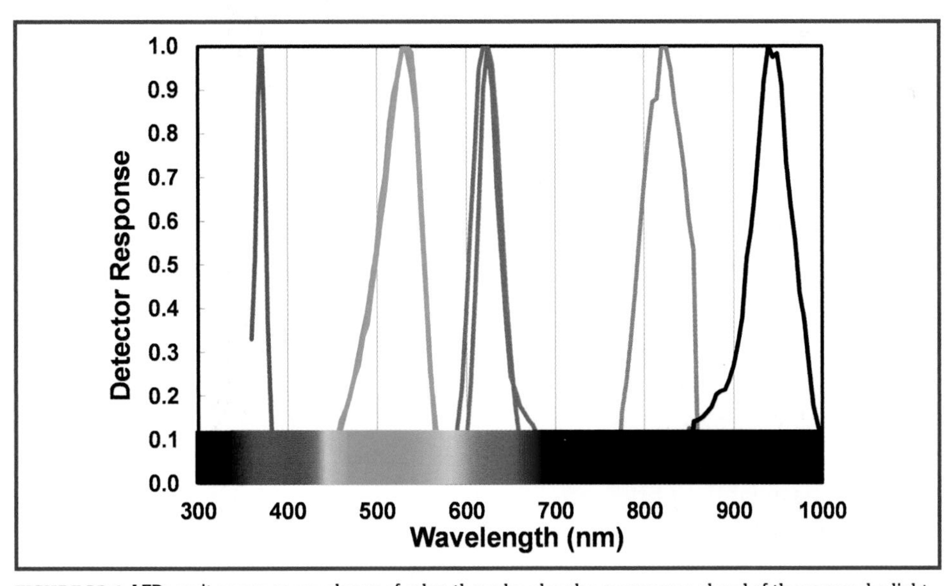

FIGURE 22-1. LEDs emit a very narrow beam of color; they also absorb a very narrow band of the same color light.

Note: This does not apply to white LEDs, which are blue-emitting LEDs coated with a phosphor that glows yellow and red when stimulated by blue from the LED. The merging of the blue, yellow, and red provides white light. Although a white LED can detect blue light, a blue LED is a much better choice.

Blue and most green LEDs are made from gallium nitride (GaN). The brightest red LEDs are made from aluminum gallium arsenide (AlGaAs). The LEDs used in near-infrared remote controllers are also AlGaAs devices; their peak emission is about 880nm and peak detection around 820nm. Older remote controllers used gallium arsenide compensated with silicon (GaAs:Si). These LEDs emit at about 940nm, which makes them ideal for detecting water vapor, but they've become very difficult to find.

In my experience, the sensitivity of red "super-bright" and AlGaAs LEDs and similar near-IR LEDS is very stable over many years of use. Green LEDS made from gallium phosphide (GaP) are also very stable. However, a blue LED made from GaN has declined in sensitivity more than any LED I have used.

Basic LED Sensor Circuits

You can substitute an LED for a standard silicon photodiode in most circuits. Just be sure to observe polarity. Also, remember that the LED isn't as sensitive as most standard photodiodes and will respond to a much narrower band of light wavelengths.

For best results, use LEDs encapsulated in clear epoxy and try a few experiments first. These will help you understand how the detection angle of an LED used as a sensor matches its emission angle when used as a light source.

Use standard couplers to attach LEDs to plastic optical fiber, or attach them directly using this method (Figure 22-2): Flatten the

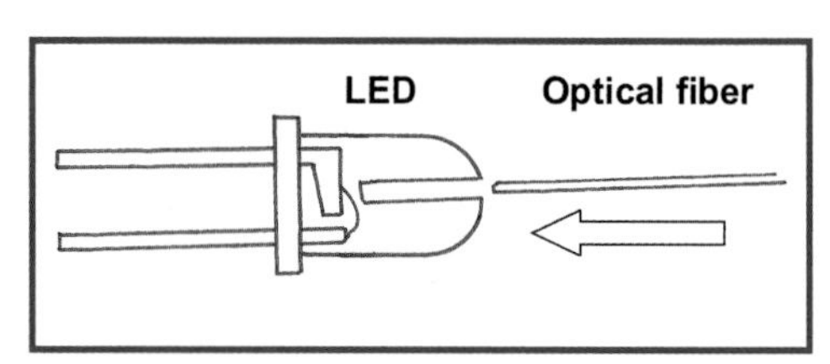

FIGURE 22-2. Use clear cyanoacrylate "super glue" to attach the fiber to the LED.

top of the LED with a file, clamp it securely, and carefully bore a small hole just above the light-emitting chip. Insert the fiber and cement it in place.

Connect the leads of a clear encapsulated red or near-IR LED to a multimeter set to indicate current. Point the LED toward the sun or a bright incandescent light, and the meter will indicate a current (Figure 22-3).

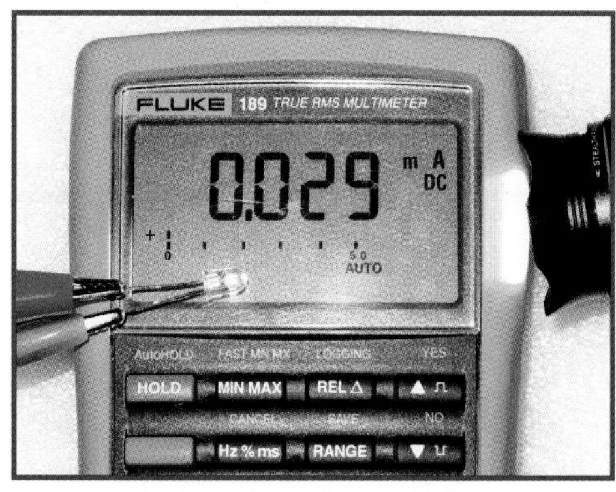

FIGURE 22-3. A brightly illuminated LED will give off a small but detectable voltage!

You can also use one LED to power a second LED. Connect the anode and cathode leads of two clear encapsulated, super-bright red LEDs. When one LED is illuminated with a bright flashlight (Figure 22-4), the second LED will give off a dim red glow. Heat-shrink tubing is placed over the glowing LED to block light from the flashlight.

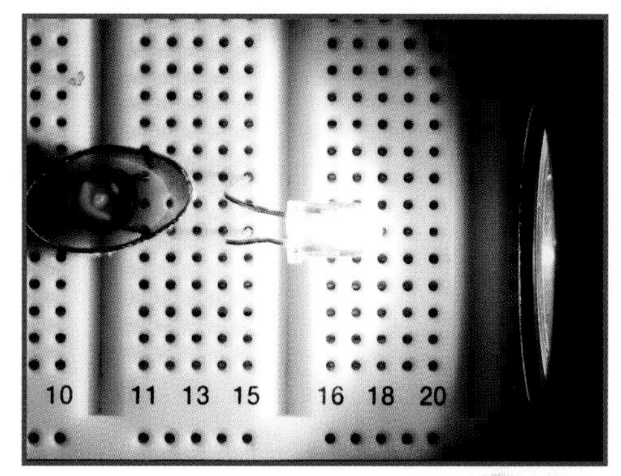

FIGURE 22-4. An LED illuminated by a white light flashlight emits a small electric current, enough to power another LED!

Figure 22-5 shows a simple circuit I often use to convert the photocurrent from an LED into a proportional voltage. The Linear Technology LT1006 single-supply op-amp (IC1) provides a voltage output that's almost perfectly linear with respect to the intensity of the incoming light. The gain or amplification equals the resistance of the feedback resistor (R1). Thus, when R1 is 1,000,000 ohms, the gain of the circuit is 1,000,000. Capacitor C1 prevents oscillation.

Many other op-amps can be substituted for the LT1006, but most of them require a dual-polarity power supply. If you use one of these, connect pin 4 directly to the negative supply. Connect pin 3 and the cathode of the

LED to ground (the junction between the minus side of the positive supply and the positive side of the minus supply).

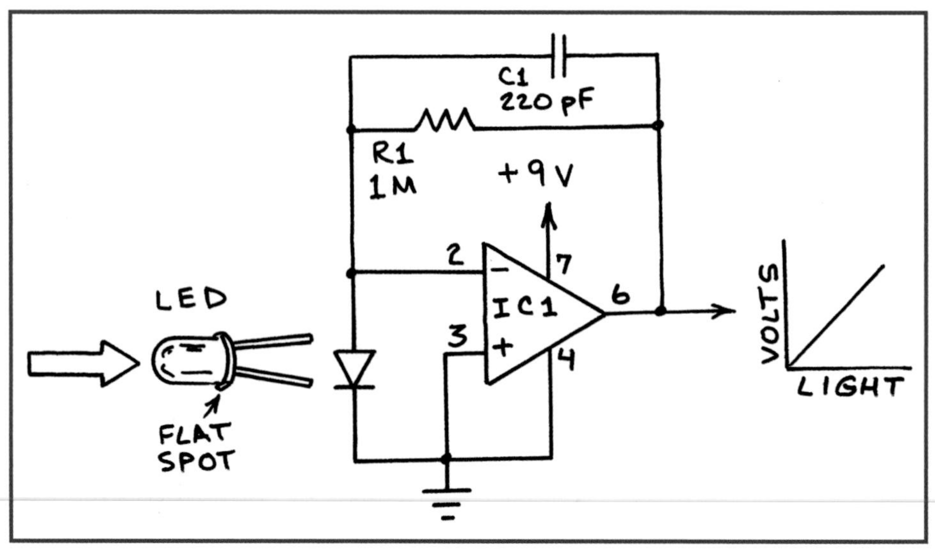

FIGURE 22-5. LEDs have a much smaller light-sensitive surface than most silicon photodiodes, so they're more likely to require amplification. Inexpensive operational amplifiers are ideal.

Going Further

The best way to come up with new applications for LEDs operated as photodiodes is to experiment with the applications I've described here. When I was doing this back in the '60s, I had no idea those simple experiments would lead to two-way communication over a single optical fiber and several kinds of instruments to measure the atmosphere that I've been using since 1989.

Use LEDs to Track Night-Launched Projectiles

23

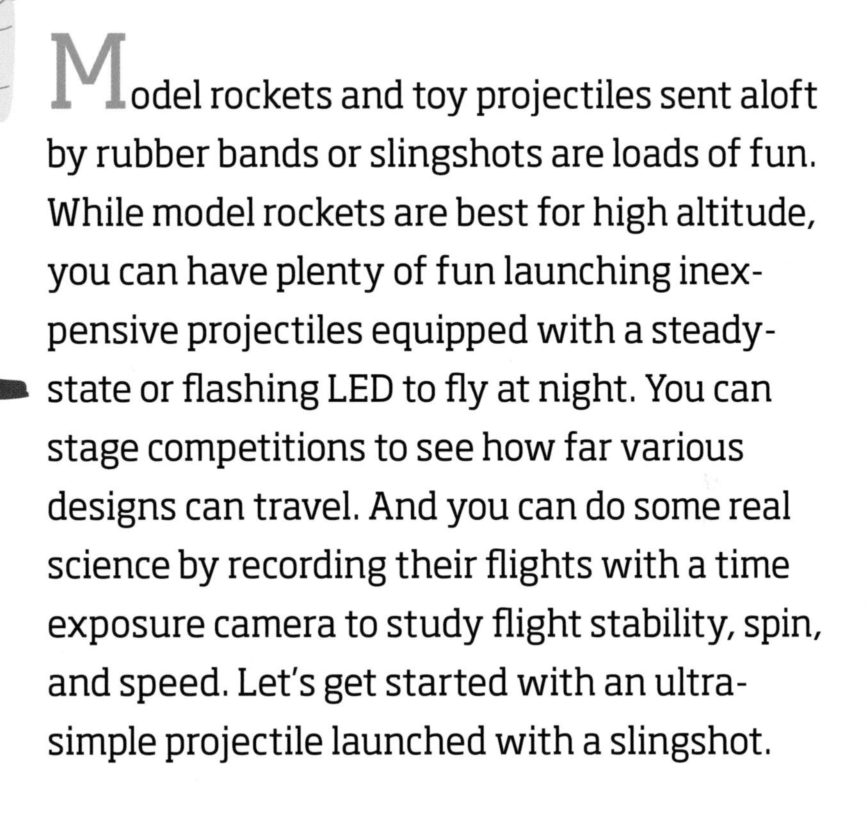

Model rockets and toy projectiles sent aloft by rubber bands or slingshots are loads of fun. While model rockets are best for high altitude, you can have plenty of fun launching inexpensive projectiles equipped with a steady-state or flashing LED to fly at night. You can stage competitions to see how far various designs can travel. And you can do some real science by recording their flights with a time exposure camera to study flight stability, spin, and speed. Let's get started with an ultra-simple projectile launched with a slingshot.

Slingshot Micro-Light Rocket

Hobby Lobby stores sell a package of a dozen miniature white LEDs installed in tiny metal cylinders equipped with a battery and rotating switch. These bullet-shaped lights are ideal for use as tiny inertia projectiles launched from a slingshot at night. They can be launched as-is or stabilized by taping one end of a broom straw, 6" long or so, to the LED fixture so that it resembles a miniature bottle rocket. Switch on an LED (or two) and place it in a slingshot pocket (Figure 23-1). If a broom straw is attached, it should point straight up so it won't hit the slingshot fork. Then fire the projectile into the air from a grassy field.

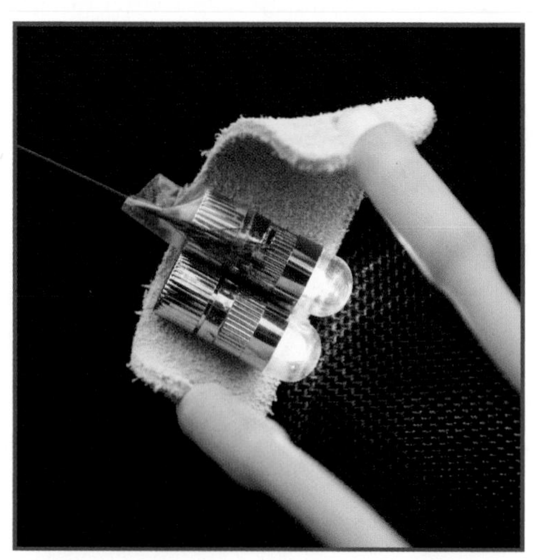

FIGURE 23-1. Fingertip-sized white LED lights.

Air resistance has very different effects on these tiny projectiles. The unmodified LED will tumble in flight, while a broom straw version will usually provide a more stable flight. However, the bare LED sometimes reaches a higher altitude than the broom straw-equipped LED. This can be seen in the time exposure in Figure 23-2. Launched simultaneously by the same slingshot, a bare LED

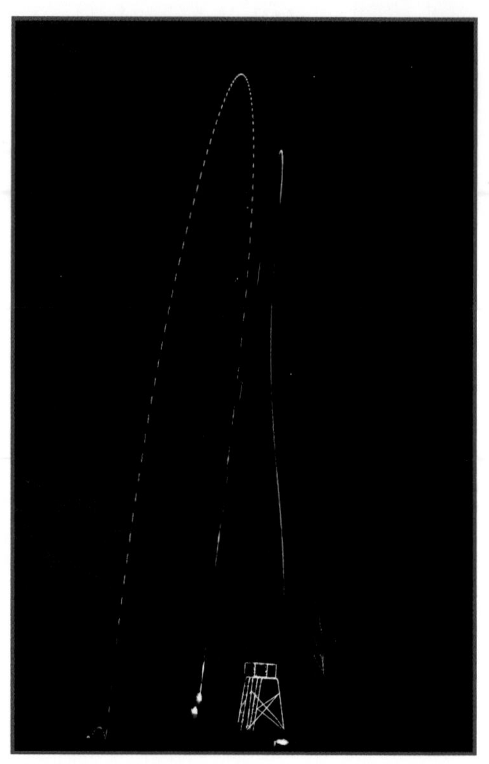

FIGURE 23-2. Time exposure of the slingshot launching of two LED projectiles.

reached 102.6 feet and a straw-stabilized LED only 90.8 feet. The altitude was determined by using the known height of the adjacent tower illuminated by a flashlight as a reference.

You can have lots of fun with these micro rockets while doing some real science. For example, what's the minimum length of the straw for stable flight? What's the maximum distance achieved by launching micro-rockets at an angle? How high can they travel? Why does the bare LED travel as high or higher than the straw-stabilized LED, and why is its tumble so regular?

You can also organize competitions to see whose micro rocket covers the most distance or lands closest to a target on the ground. Paint the LEDs with different colors and leave them glowing in the field until all the competitors have completed their flights.

More Powerful Inertia Rockets

During a recent visit to the White Sands Missile Range Museum, I purchased a package of three Sky Rockets made by Monkey Business Sports. These "High Flying, Hand Launched, Foam Rockets" are sent on their way by a handheld elastic launcher. If you flick your wrist forward while releasing the rocket, it can reach an altitude of 300 feet.

It's easy to equip a Sky Rocket with a tracking light. Tape the leads of a white LED across opposite sides of a lithium coin cell so that the lead on the flat side of the LED is on the negative side of the cell, and tape this assembly to the back end of a rocket (Figure 23-3). (Optionally, you can add a simple on-off switch using a paper tab, as shown in Step 3 at http://makezine.com/projects/extreme-led-throwies/.)

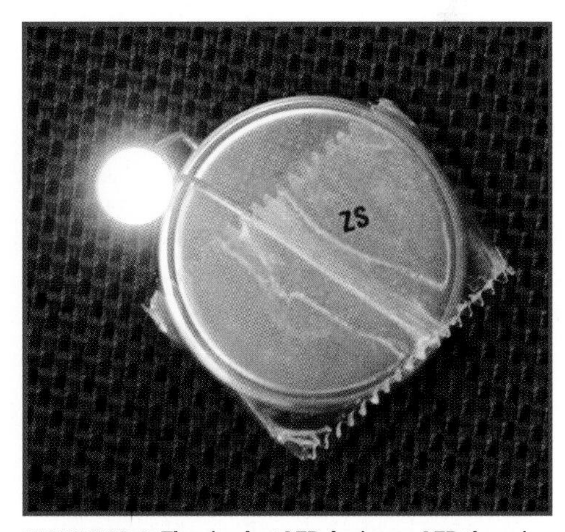

FIGURE 23-3. The simplest LED device, an LED throwie, made by taping an LED to a coin cell battery.

To study roll rate, insert a Hobby Lobby LED into a hole made in the side of the rocket at its center of gravity (balance point) and secure it in place with clear tape (Figure 23-4). The time exposure in Figure 23-6 shows a typical flight, with the rocket's rotation clearly indicated by streaks of light. The streaks are longest during the early stage of flight when the missile is moving fastest. The ground streaks in the photo are from my flashlight as I walked to the launch site.

For a tracking light that flashes, I've not found anything better than Coghlan's Brite Strike APALS, an ultra-thin, lightweight rectangle measuring 1"×2" that is easily stuck to the side of a rocket and secured with clear tape. The time exposure in Figure 23-6 shows the apogee of a flight neatly outlined by this red flasher.

Photographing Night-Launched Missiles

Digital cameras are ideal for recording the flight path of night-launched rockets and inertia projectiles equipped with a tracking

FIGURE 23-4. LED throwie taped to a Sky Rocket.

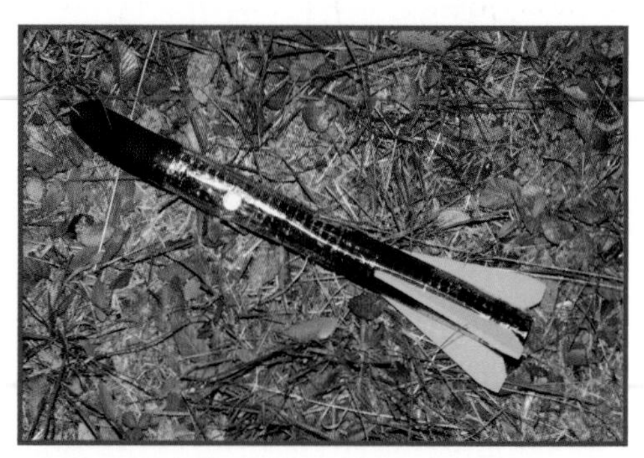

FIGURE 23-5. A Sky Rocket with attached LED throwie.

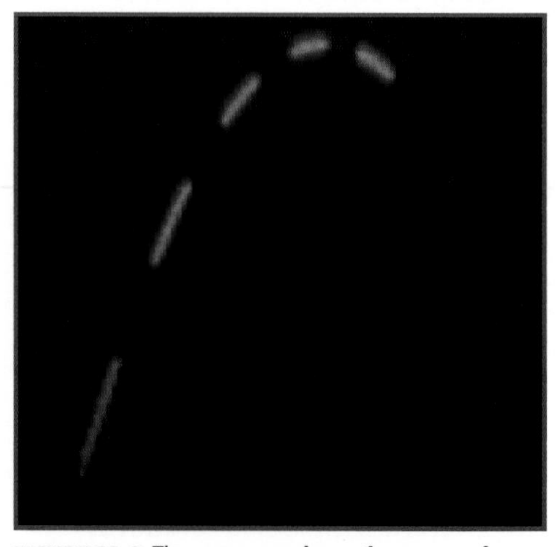

FIGURE 23-6. Time exposure shows the apogee of a flight neatly outlined by a red LED flasher.

light. The flight paths in Figures 23-2 and 23-6 were photographed with a Canon 7D set for an ISO of 6400. The lens was a 20mm-40mm wide angle. The camera was mounted on a tripod 150 feet from the launch site.

For early flights I set the exposure time to 30 seconds, tripped the shutter, ran to the launch site, and launched the rocket using an elastic-powered hand launcher or slingshot. After falling down during one night session, I began triggering the camera with a radio-controlled actuator, which greatly simplified the launch protocol. After the camera was set for a 9-second exposure, I walked to the launch site, pressed the transmitter button, and launched the rocket.

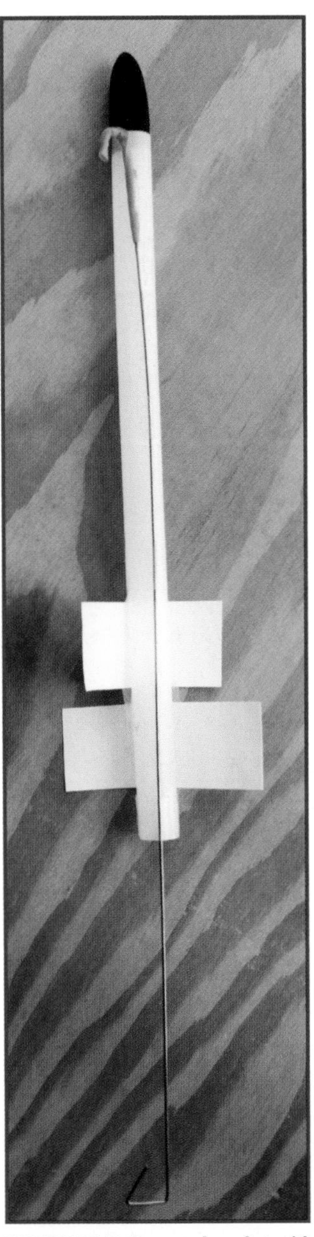

FIGURE 23-7. Homemade rocket with aper fins.

The altitude reached by your rockets can be determined by including an illuminated object of known height in each photo. Dark nights with few or no clouds are best.

Going Further

The sky's the limit with this project. You can easily build your own inertia projectiles using paper or plastic tubes equipped with paper fins and a foam nose from a toy missile or a miniature paint roller. I've built and launched simple rockets using a homemade launcher I made by attaching a loop of neoprene tubing to a handle. Night tracking would also be fun with water rockets (http://makezine.com/projects/make-05/soda-bottle-rocket/), compressed-air rockets (http://makezine.com/projects/high-pressure-foam-rocket/), and other DIY rockets and projectiles.

Safety Precautions

As with any project involving projectiles, use common sense. Children should be supervised. Rockets should never be launched toward people or structures, and should always be launched from a field. Inspect the field during daylight to check for potential safety hazards.

Wear clear safety glasses when launching elastically propelled rockets. During night launches, all participants should have a flashlight. I wear a headlamp to keep my hands free.

Connecting Fibers to LEDs and Sensors

Optical fibers are ultra-clear strands of plastic, glass, or silica consisting of a central core surrounded by a cladding and a protective coating. Light injected into the core of a fiber remains trapped until emerging from the opposite end. This lets you transmit light point-to-point with very little loss, and even bend it around corners. The light stays in the core because the cladding has a slightly higher index of refraction than the core.

Silica optical fibers are primarily used to transmit high-bandwidth data over long distances. Inexpensive plastic fibers are widely used in sensors, illuminators, and toys. They're also used to couple light to photodiodes in environments that require electrical isolation or protection from the elements or a corrosive environment. Plastic fibers are also used to illuminate displays and to send light through openings too small for a flashlight.

Optical fiber couplers for various LEDs and light sensors are commercially available, but you can skip the connector and simply connect silica and plastic fibers directly to LEDs and sensors. For the examples described here, I used LEDs encapsulated in standard 5mm clear epoxy packages, and 2.2mm-diameter plastic fiber with a 1mm core and a black polyethylene jacket. I used fiber from Jameco Electronics, but many kinds of fiber are available from other online sources, including eBay. The methods described here can be adapted for use with most of these fibers.

Preparing the Fiber Ends

For most applications, you can obtain best results with a fiber that has flat, smoothly cut or polished ends. A simple way to achieve this with plastic fiber is to place the end of the fiber on a wood surface and slice off a few millimeters by pressing a sharp hobby knife blade straight down into the fiber. Press the cut end of the fiber against 200-grit

FIGURE 24-1. The roughened edges of the fiber's jacket can be trimmed away with a hobby knife.

sandpaper on a flat surface, and rotate the fiber in a dozen or so circles across the paper. Follow this with 400- or 600- grit paper. A fiber end polished in this fashion is shown in Figure 24-1.

Materials

- Plastic optical fiber, 2.2mm
- LEDs, phototransistors, or photodiodes, 5mm epoxy encapsulated
- Heat-shrink tubing, 6mm or 8mm
- Plastic cement or cyanoacrylate (CA) glue, aka super glue

Tools

- Hobby knife
- Sandpaper: 200 grit and 400 or 600 grit
- Magnifier, 10x
- Drill or high-speed rotary tool Dremel 7700 or similar
- Drill bits: 3/64" and 7/64"
- Vise or clamp
- Heat gun, butane lighter, or other source of heat

Heat-Shrink Tube Connection

Heat-shrink tubing provides the simplest way to connect optical fibers to LEDs and sensors. This method is not necessarily practical for long-term use, especially outdoors, but it works well for basic experiments and demonstrations with 2.2mm fiber. For best results, use 6mm- or 8mm-diameter heat-shrink tubing and a 5mm LED or sensor.

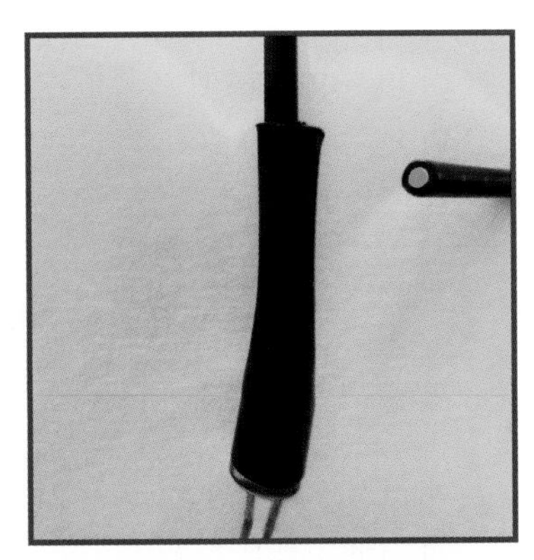

FIGURE 24-2. You can keep the fiber in place by applying some adhesive, or you can allow the fiber to be removable.

Slip a 1" length of tubing over the LED or sensor, and warm the tubing with a heat gun until it holds the LED or sensor tightly in place. Slip the end of the fiber into the open end of the tubing and continue warming. Depending on the diameter of the tubing, the fiber will be anchored in place or it can be slipped in and out of the tubing (Figure 24-2).

Permanent Direct Connection

You can make a stronger coupling by cementing the fiber into a hole bored into the end of the LED or sensor's epoxy capsule. Cement the jacketed end of the fiber, or remove a section of jacket and cement only the fiber itself. An inch or so of heat-shrink over the junction will finish the job. Follow these steps:

1. Secure the leads of the LED or sensor in a vise or a DIY clamp made from a clothespin and a large binder clip.

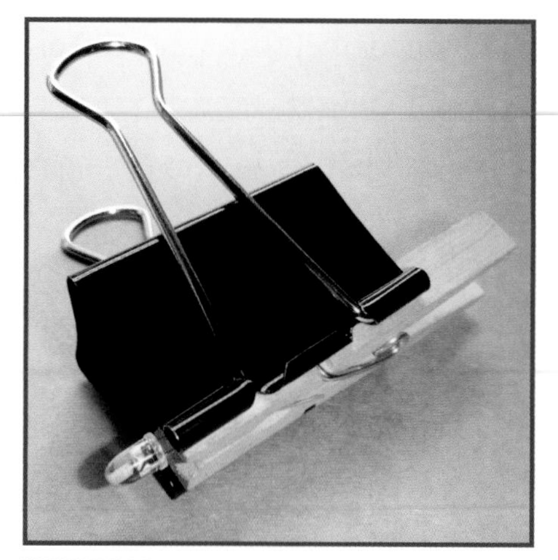

FIGURE 24-3

2. Use a fine Sharpie to draw a plus sign (+) directly over the end of the LED. Insert a 3/64" bit into the chuck of a drill. I prefer the handheld, battery-powered Dremel 7700. Lightly touch the spinning bit to the center of the plus sign. Let the drill do the work as you guide it straight into the device while applying very gentle pressure.

FIGURE 24-4

3. Carefully bore the hole to just above the tiny wire(s) that make contact with the light-emitting or -sensing chip.

   The hole produced by a 3/64" bit should accept the 1mm bare fiber core. If you're connecting jacketed 2.2mm fiber, carefully enlarge the hole with a 7/64" bit.

4. Use compressed air to blow away any chips in the hole. Be sure any connection leads on the top of the chip are undamaged.

5. Insert the polished end of the fiber into the hole, then secure it with cyanoacrylate or other plastic adhesive, and let it dry. Figure 24-6 shows a fiber with a bare end inserted into a blue LED, and Figure 24-7 a jacketed fiber inserted into a white LED.

6. Insert a suitable length of dark heat-shrink tubing over the LED or sensor and 1" or so of the fiber, and warm it to secure the tubing in place.

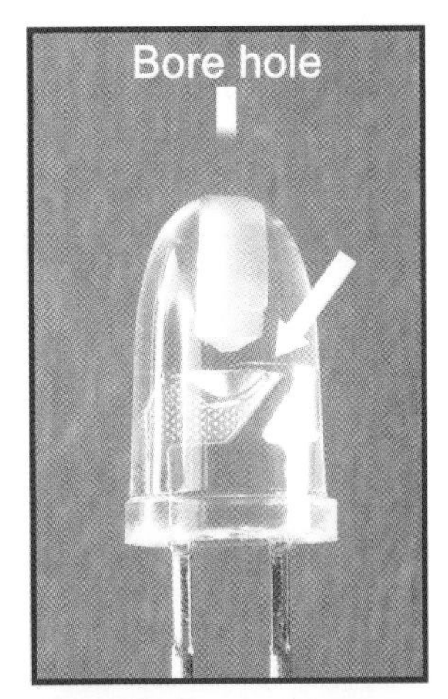

FIGURE 24-5

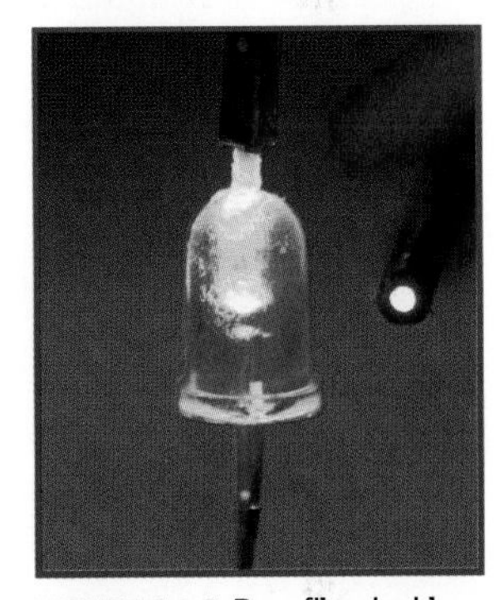

FIGURE 24-6. Bare fiber inside a blue LED.

If you use the fiber with a light sensor, paint the exposed base of the sensor with black enamel to block external light. You may need additional blockage because infrared wavelengths penetrate black paint.

Going Further

You can start using your connected fibers right away for illuminating your projects, props, and models, or for photography. I had fun experimenting with "light painting" by making 9-second time exposures (Figure 24-8). Pulse the LED

FIGURE 24-7. Jacketed fiber inside a white LED

to create dashed instead of continuous lines in the images, and use multiple fibers to add more colors.

In the next chapter I'll show you how to use fibers connected to LEDs and phototransistors as sensitive sensors.

Meanwhile, you can expand on the methods described here by designing your own connections. Use a ballpoint pen housing to make a handheld optical fiber pixel probe or micro-light source. Or consider 3D printing your own custom-designed connectors and fixtures.

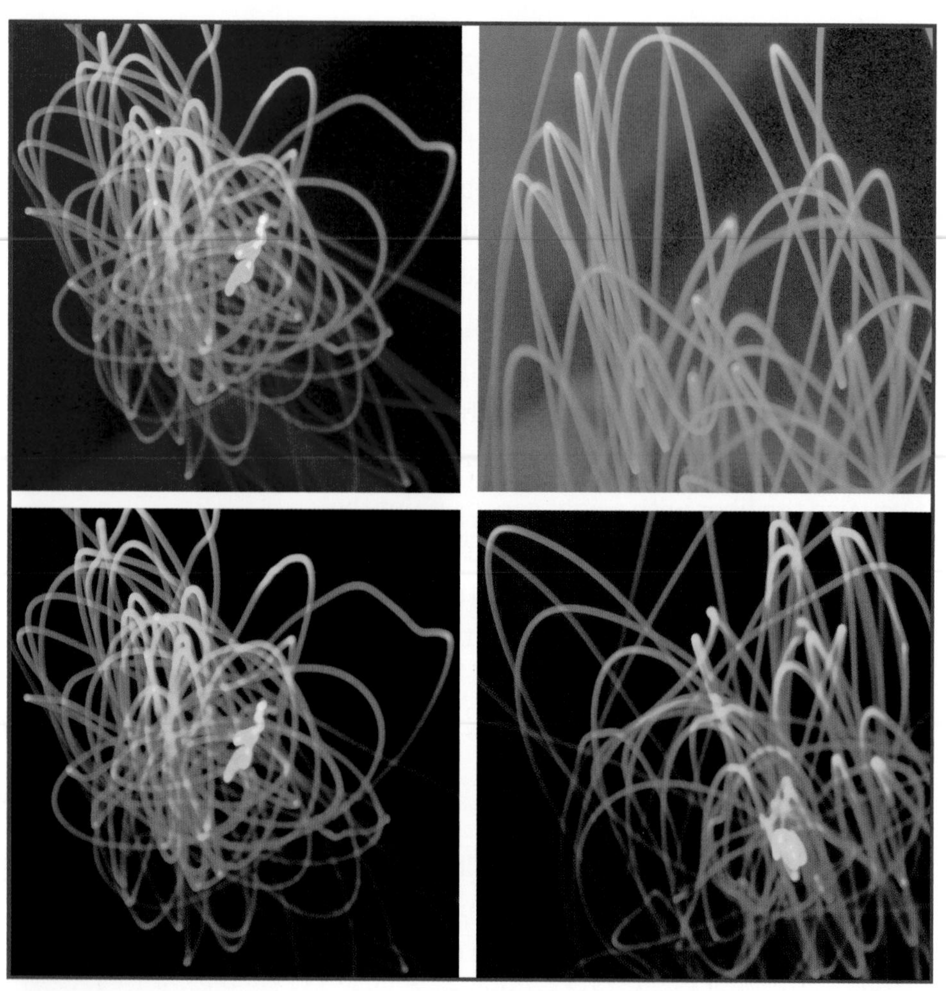

FIGURE 24-8. Pulse the LED to create dashed instead of continuous lines in the images, and use multiple fibers to add more colors.

25 Transform Things into Sounds with the PunkPAC

Since Punk Science is the theme of the issue of *Make* in which this column originally appeared, let's explore a new twist on electronic tone generators like the popular Atari Punk Console (APC). Why not hack the APC so that its tones can be controlled by light?

We'll do this by interfacing an optical fiber to a basic light-sensitive oscillator and an APC to create a Pixel-to-Audio Converter (PAC), which you could also call a Photon- or Photo-to-Audio Converter.

I call the resulting circuit a PunkPAC. It can be used to transform optical patterns formed by photos, fabrics, and even tree rings into variable-frequency tones for making music (or noise). It could easily be used for interactive concerts and museum exhibits.

Atari Punk Console

The 555 timer is the most popular of the many integrated circuits designed by legendary engineer Hans R. Camenzind. More than 30 years ago I described how to make a simple sound synthesizer by connecting the output of a 555 oscillator to the voltage control pin of a second 555 oscillator. This provided a stepped-tone generator in which slow pulses from the first oscillator altered an audio frequency tone from the second oscillator. The pulse repetition rate of each oscillator was controlled by a potentiometer, and appealing tone sequences could be created by adjusting either or both of them.

Electronic music experimenters began experimenting with the circuit, which was dubbed the Atari Punk Console (APC) by Kaustic Machines (http://compiler.kaustic.net/machines/apc.html) because it makes sounds similar to the classic Atari 2600 video game console. The APC remains popular and even has its own Wikipedia page. Googling "Punk Console" yields more than 400,000 hits and over 1,000 video clips on YouTube.

Controlling the Punk Console with Light

The pots that control the frequency of a 555 tone generator and the APC can be replaced with light-sensitive photoresistors. This lets you "play" these circuits by waving your hands between the photoresistor(s) and a light source. Adding a DIY optical-fiber probe provides much finer control and the ability to transform natural and manmade patterns into distinctive tone arrangements.

Build an Optical Fiber Interface

You can make a simple but sturdy optical PAC probe from a jacketed plastic optical fiber (Jameco Electronics part #171252 or similar), a miniature CdS photoresistor with a dark resistance of 10MΩ or more (Jameco #202391 or similar), and a plastic applicator tip (#SIGSH10000 from hobby shops), as shown in Figure 25-1. Follow these steps to make it:

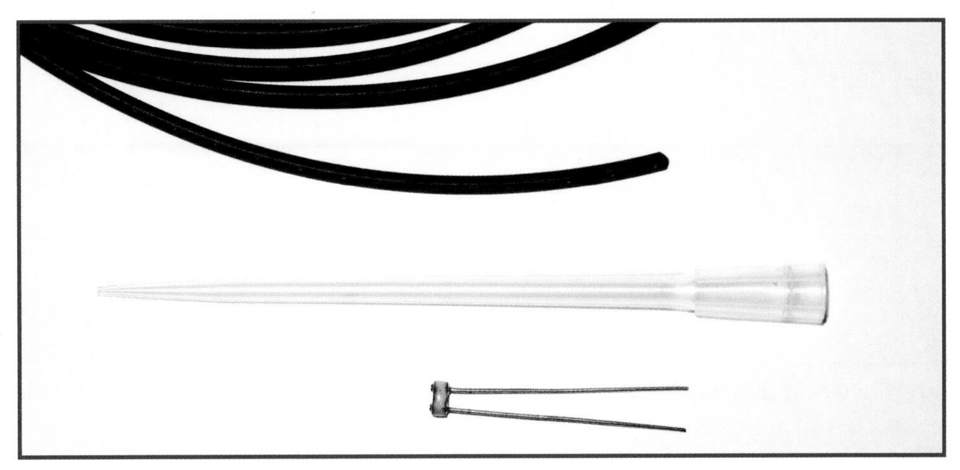

FIGURE 25-1. Components for a Pixel-to-Audio Converter (PAC) light probe.

1. The optical fiber comes in a 10' coil. Select the end with the cleanest cut. If both ends are rough, slice off one end with a sharp hobby knife. Make a second cut 10" up the fiber.
2. Use the hobby knife to slice through the long applicator tip 44mm from its small opening. The 10" optical fiber should fit through the hole in the tube. Recut if necessary.
3. Bend the leads of the photoresistor so they fit over the large end of the applicator.
4. Use long-nose pliers to form a small U shape in one end of each of a pair of connection wires that fit the holes in a solderless breadboard. Crimp a wire U around the bend in one of the photoresistor leads and solder it in place (Figure 25-2). Repeat with the second connection wire and photoresistor lead.

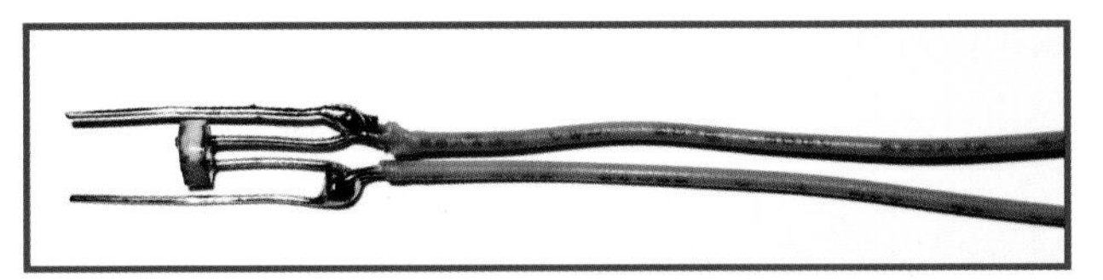

FIGURE 25-2. Breadboard-style connection leads are soldered to the photoresistor leads.

5. Insert the photoresistor into the large end of the applicator (Figure 25-3), and hold it in place while you push the optical fiber into the small end until it touches the photoresistor.

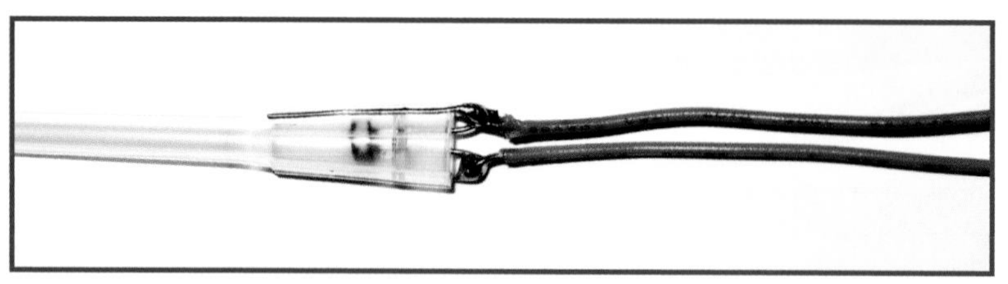

FIGURE 25-3. The photoresistor is inserted into the large end of the applicator tip.

6. Wrap black electrical tape over the photoresistor leads to hold them in place and block stray light (Figure 25-4). Or slide a 1½" section of black heat-shrink tubing over the photo-resistor and warm it with a hair dryer.

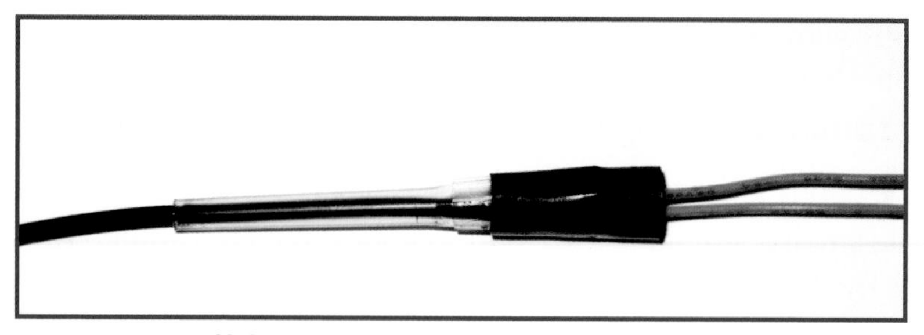

FIGURE 25-4. Assembled PAC light probe. The backside of the probe should be blocked from external light.

PunkPAC: A Light-Controlled Tone Generator

Figure 25-5 shows a basic PunkPAC circuit made from a 555 tone generator. You can easily build it on a solderless breadboard. Insert the photoresistor leads from the PAC probe into the holes adjacent to pins 7 and 8 of the 555, check your wiring, and connect a 9V battery.

Switch off any lights, and the speaker should emit clicks or a tone. The frequency of the tone should rapidly increase when the optical fiber is pointed toward light. Increase the value of C1 to reduce the tone frequency. Try listening to the tone patterns and sequences that are produced when you scan the fiber-optic probe across these words on a printed page or a computer monitor.

We'll return to more applications shortly, but first let's build an Atari Punk Console.

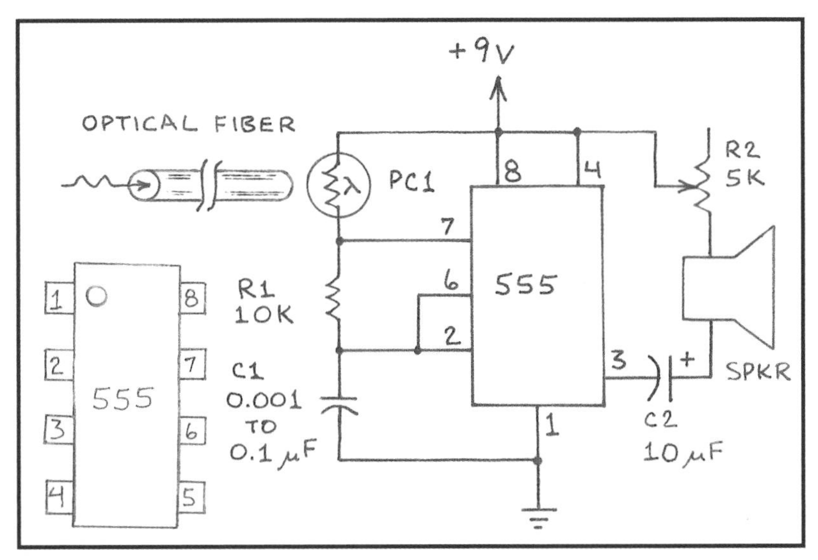

FIGURE 25-5. Basic 555 timer light-to-tone generator circuit.

A Stepped-Tone (APC) PunkPAC

Figure 25-6 shows an Atari Punk Console circuit with a PAC probe. You can build the circuit on a solderless breadboard or assemble it from a kit version. Either way, carefully check the wiring before connecting a 9V battery.

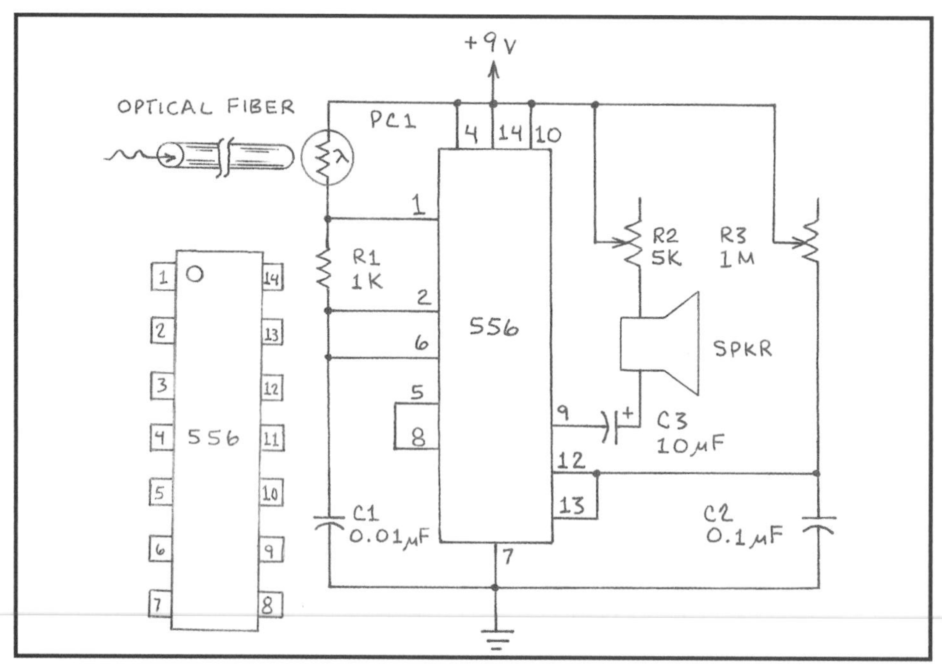

FIGURE 25-6. PunkPAC 556 stepped-tone circuit for use with a PAC light probe.

The unique stepped-tone output from the APC differentiates it from the basic 555 tone generator in Figure 25-5. Instead of a continuously variable tone controlled by light entering the PAC probe, the APC tone changes in distinct steps. The values for C1 and C2 in Figure 25-6 are very flexible. Experiment with different values to arrive at the best effects.

Applications for PunkPAC Circuits

Years is a clever creation by multimedia artist Bartholomäus Traubeck in which a record player spins a thin cross-section from a tree (http://traubeck.com/years/). A light sensor detects changes in the annual rings and feeds the data to a computer that responds with piano notes.

FIGURE 25-7. Bartholomäus Traubek's artwork *Years*.

Since the rings aren't perfectly concentric, the sensor doesn't track individual rings but moves back and forth across them. The best musical bursts occur when it sees blemishes caused by age or disease (Figure 25-7).

PunkPACs allow you to manually transform tree rings into variable or stepped audio tones more faithfully than the *Years* system.

Illuminate a smooth cross section with a lamp and place the PAC probe over the wood (Figure 25-8). Not much happens when you trace an individual ring. There's much more tone variation when you slide the probe across the rings to create an undulating sequence. Light-tinted early wood creates a higher frequency than darker late wood, so high-contrast rings work well. You can "play" a wide ring from a wet year, a thin drought ring, or the ring that grew the year you were born.

FIGURE 25-8. Using the PAC light probe to scan the rings of a bois d'arc tree felled by a flood.

For better results, scan the probe across images of tree rings on a smartphone, tablet, or computer screen. For best results, convert them to high-contrast black and white, as in Figure 25-9. For variety, try images of feathers, rocks, bark, fabric, and wallpaper.

FIGURE 25-9. Using the PAC light probe to scan a high-contrast image of tree rings on a smartphone display.

Going Further

The PunkPAC lets you extract distinctive musical tones from everyday things, and its square-wave output can be processed by a microcontroller to trigger programmed notes emitted by various musical instruments.

Imagine a concert where you email images or videos to the conductor, who scans them with a PunkPAC probe to embellish the music. Mine would be a video of sunrise in Hawaii (see www.youtube.com/user/fmims for a sample).

Making Synthesized Music from Your Data

26

Do you enjoy the sounds that wind chimes extract from a soft breeze? How about the gentle splashing of raindrops, the soothing sound of falling water, or the roar of surf?

These appealing natural sounds provide only a hint of the vast range of musical compositions hidden away in many kinds of data. Lately I've been having lots of fun transforming data I've collected into natural music and posting the results on www.youtube.com/user/fmims. Various methods for making sounds from data are available. Let's use them to convert data into music.

Using Mathematica to Create Music

George Hrabovsky is an amateur physicist who uses Wolfram Mathematica software (www.wolfram.com/mathematica/) in his theoretical research. His praise of Mathematica was so persuasive that I eventually bought the program, and it's where I first went when trying to transform data into music.

Among Mathematica's astonishing range of features is the ability to convert numbers into synthesized musical notes representing a variety of instruments. Mathematica's Music Package can convert data into representative audio frequencies and much more. If you're into programming, it's a highly flexible tool for transforming data into music.

MusicAlgorithms

Jonathan Middleton is Assistant Professor of Theory and Composition in the music department at Eastern Washington University, where he teaches composition, orchestration, and computer music. While exploring ways to transform data into music, I discovered Dr. Middleton's MusicAlgorithms website (developed with assistance from Andrew Cobb, Michael Henry, Robert Lyon, and Ian Siemer, with sponsorship from the Northwest Academic Computing Consortium).

The homepage states that "Here, the algorithmic process is used in a creative context so that users can convert sequences of numbers into sounds." That single sentence hooked me into the MusicAlgorithms site for a week while I transformed some of my data into an amazing variety of intriguing musical "compositions."

How to Use MusicAlgorithms

MusicAlgorithms requires a Java-enabled computer. Transforming a string of numbers into music is simple; you can either type or paste a series of numbers into the program. Here's a quick way to learn to use the site:

1. From the homepage (http://musicalgorithms.ewu.edu/), click the Compose button. Then click "Import your own numbers" to enter the data input page.
2. In the Algorithm box, enter into window A the numbers 1 through 10 (press the Enter key after each number).
3. Ignore checkboxes B, C, and D, and click the Get Algorithm Output button.
4. In the Pitch box, click the Scale Values button to normalize the numbers 1 to 10 that you entered in window A into the piano scale of 0 to 88 (1 = 0, 2 = 9, 3 = 19, ... 10 = 88). These numbers will appear in the adjacent Derived Pitch Values window.
5. Skip the Duration box (for now). In the Compose box, click the Play button.
6. A MIDI Player window will open, showing a piano keyboard over buttons for Step and Play and options for Volume, Tempo, and Instrument (Figure 26-1).

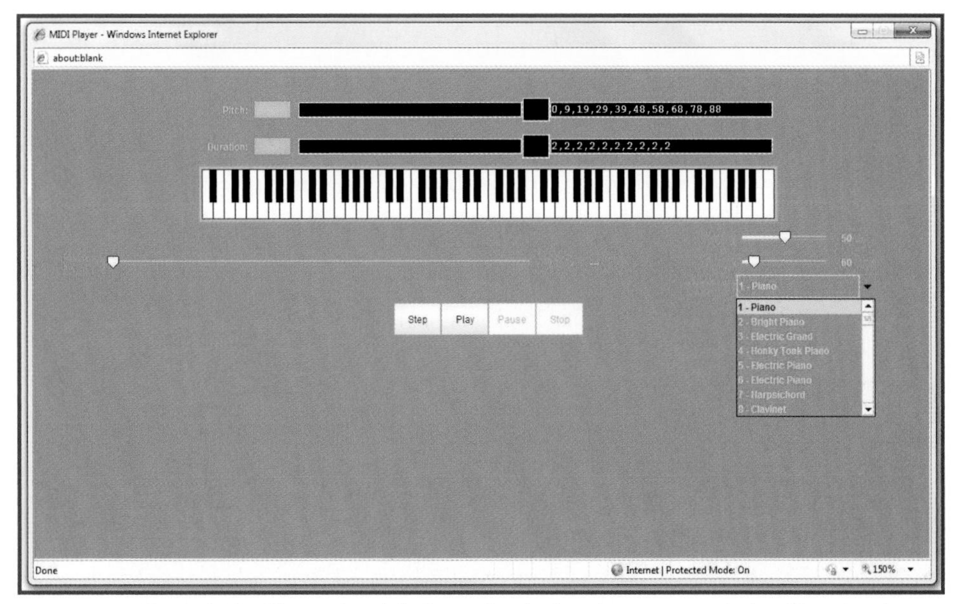

FIGURE 26-1. The MusicAlgorithms web tool displays a piano keyboard over a range of selectable options for instrument selection, tempo, and volume. A pointer on the scale below the keyboard keeps track of progress.

Click the Play button to hear the 10 notes you have composed. Keep playing these notes while using the sliders to adjust the volume and tempo. Then let the fun begin by selecting from the pull-down menu of 128 synthesized instruments and sounds. Soon you'll be ready to compose music from real data.

Finding Data for MusicAlgorithms

If you're an amateur weather watcher, you probably have plenty of numbers to transform into music. For example, MusicAlgorithms will convert a year of your daily minimum and maximum temperatures into a remarkable audio experience that will provide an entirely new way to appreciate your data. If you have no scientific data, try converting your daily expenses or bank balance into music. You might be surprised by what you hear.

A goldmine of data is scattered across the web. For example, my local National Weather Service station near San Antonio, Texas, provides monthly and annual precipitation data since 1871 and temperature since 1885. Converting these data into music provides an entirely new way to better appreciate seasonal temperature cycles and even cold fronts, El Niños, and droughts.

The U.S. Geological Survey provides data on stream flow. Many NASA and NOAA sites are filled with data. Other data sources include the U.S. census, stock market statistics, commodity prices, grocery store price lists, traffic counts on major highways, and so forth.

Sample MusicAlgorithms

I've posted several videos of MusicAlgorithms based on my data. These will give you a good idea of the amazing variety of sounds you can produce from data that ordinarily are depicted only as dots, lines, or bars on charts.

- One year of solar noon UV-B data. Since 1988, I've measured the sun's ultraviolet radiation from a field in Texas at solar noon on days when clouds didn't block the sun. In this YouTube clip (www.youtube.com/watch?v=VsRCrh6XWog), the UV-B intensity at noon on each of the 170 days during 2010 in which the UV-B could be measured (Figure 26-2), is transformed into representative musical notes.

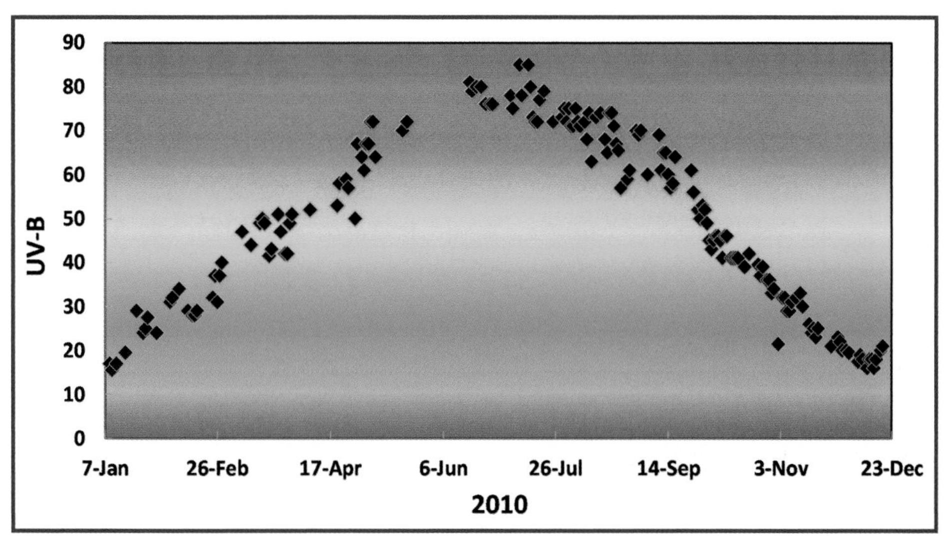

FIGURE 26-2. Weather data, like these solar UV measurements, are easily converted into representative sequences of musical tones.

Each musical note is accompanied by a 360° fisheye image of the sky made when the UV-B was measured. Low UV-B levels during winter are indicated by low pitch tones, and the high UV-B levels during summer by high pitch tones. Variations in the steady increase and then decrease in UV-B during the year are caused by clouds near the sun, haze, and changes in the ozone layer.

- The cosmic ray background count. In this video (www.youtube.com/watch?v=bAKdaYumlq4), the cosmic ray background count on a flight from San Antonio, Texas, to Zurich, Switzerland (Figure 26-3), is transformed into an audio composition in which the frequency of tones represents altitude. A typical Geiger counter measures around 11 counts per minute (CPM) at the ground and several hundred CPM at altitudes of 35,000 feet or more.

▸ Tree rings to symphonic strings. MusicAlgorithms can convert the widths of annual growth rings in trees into a tune in which wide rings (from wet years) have a higher pitch than thin rings (from dry years). This composition uses ring data from a tree at my place downed by a flood in 2010 (Figure 26-4).

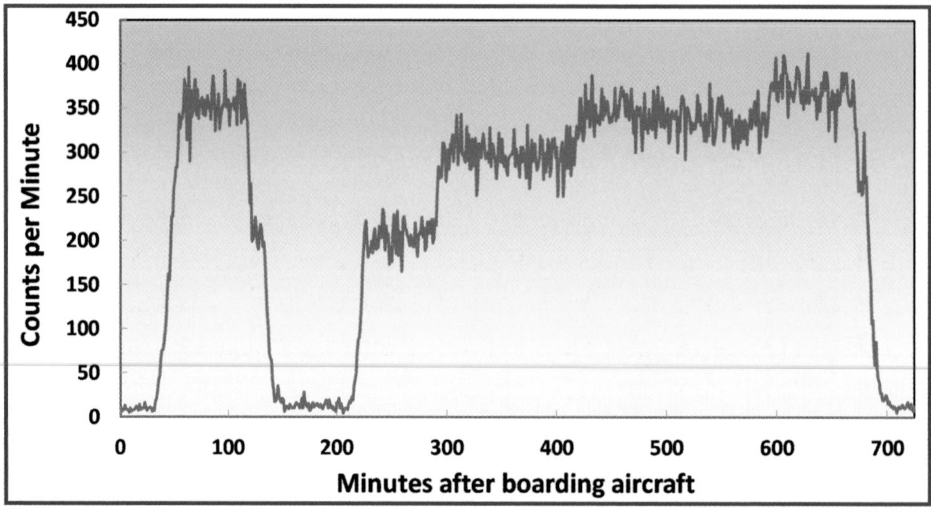

FIGURE 26-3. The cosmic ray background count measured by a Geiger counter increases with altitude. You can hear the altitude changes of an aircraft flying from San Antonio to Zurich as distinct changes in pitch, proportional to altitude.

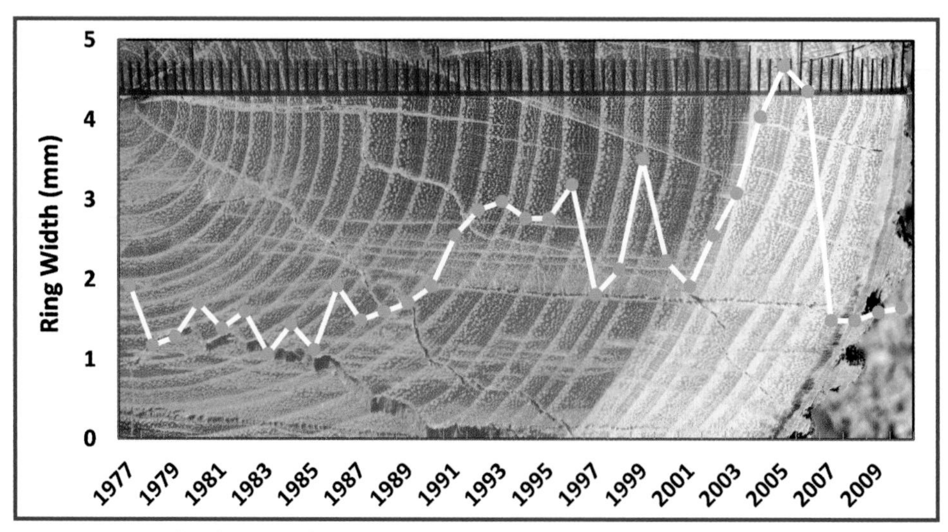

FIGURE 26-4. Decades of tree growth can be converted into music based on the precipitation-modulated width of annual growth rings. The bois d'arc tree shown here began growth in 1977.

A tree produces one growth ring each year. The light-colored spring growth is called early wood, the darker summer-fall growth is late wood. In this video (youtu.be/l2g3scrcg20), the width of the early and late wood in each ring was measured and then played in this sequence: early wood note/late wood note/sum of early and late wood notes/rest interval to separate the rings. This clip also includes the data plot I used to form the composition.

Going Further

I've simply transformed strings of numbers into music. MusicAlgorithms lets you do much more. You can select from a variety of mathematical functions and then enter the pitch range and the duration of each tone. These functions include the mathematical constants pi (π), phi (φ), and e, exponents, the Fibonacci sequence, Pascal's triangle, Markov chains, and even a chaos algorithm and DNA sequences. If you're interested in synthesized music, you can't go wrong exploring these features. Just block out some time. MusicAlgorithms is the most addictive website I've ever visited.

Startups:
Origins of the PC Revolution

Today's smartphones and tablet, laptop and desktop computers all trace their ancestry to the arrival of the hobby computer era of the 1970s.

After Intel introduced its 8008 microprocessor, in 1972, several individuals and teams began using the new chip to build DIY computers. But these computers made little progress due to the 8008's limited capabilities.

The computer revolution was jump-started in 1975 when MITS, Inc., a small electronics company that I co-founded in Albuquerque, New Mexico, announced the Altair 8800, a kit computer designed around Intel's new and powerful 8080 microprocessor.

Many books have been written about what happened next, and *Idea Man* (Portfolio/Penguin), a new memoir by Microsoft co-founder Paul Allen, shines a spotlight on many details that were previously known only to insiders. Whatever your favorite kind of computing device or operating system, *Idea Man* is a book well worth reading, especially if you have entrepreneurial aspirations.

The story begins at Out of Town News in Cambridge's Harvard Square on a snowy December afternoon in 1974. Allen visited the newsstand each month to check out the latest issues of *Radio-Electronics, Popular Science,* and similar magazines.

When he saw the January 1975 issue of *Popular Electronics,* it stopped him in his tracks. Emblazoned on the cover was a photograph of the Altair 8800 microcomputer.

The blurb over the photo read:

PROJECT BREAKTHROUGH! World's First Minicomputer Kit to Rival Commercial Models ... "ALTAIR 8800" SAVE OVER $1,000

Allen opened the magazine and found complete construction plans for the Altair 8800, which was available as a kit ($439) or fully assembled ($621). He noticed that the core of the Altair was Intel's powerful new 8080 microprocessor, the successor to the 8008. He paid 75 cents for the magazine and hurriedly strode almost a mile to Harvard's Currier House, where sophomore Bill Gates resided.

Gates shared Allen's enthusiasm for the Altair. Both had become expert assembly language programmers in high school, and they decided to contact Altair developer Ed Roberts, who headed MITS, Inc. Their plan was simple: offer Roberts a version of the BASIC language that would run on the Altair.

After eight grueling weeks of programming, Allen flew to Albuquerque with a paper tape, punched with their new BASIC. The code ran fine when simulated on a PDP-10 minicomputer at Harvard, but would it work with an Altair? While Roberts watched, Allen carefully entered into the Altair's front panel toggle switches the code he'd written on the airplane to enable the

Altair to load the BASIC from the Teletype terminal connected to the computer. The paper tape reader then loaded the BASIC into the Altair's memory. When Allen typed PRINT 2+2, the Teletype immediately printed 4.

Roberts was amazed. So was Allen, though he didn't let on. Soon Roberts hired Allen, and later that year Gates joined him in Albuquerque. There, Allen and Gates formed a partnership that they initially called Micro-Soft.

Idea Man

Allen tells what happened next in *Idea Man*, a detailed and appropriately technical account of the origin and early history of Microsoft. It's much more than a book about microcomputer history and Allen's life as a billionaire, for it's packed between the lines with tips for aspiring entrepreneurs, designers, programmers, and makers with revolutionary ideas.

Idea Man has attracted considerable attention in the media world because of its candid revelations about friction between Allen and Gates and what Allen describes as Gates' efforts to reduce Allen's stake in Microsoft.

The shouting matches he describes closely parallel what Roberts and others told me over the years. Some believe that dredging up these old stories is sour grapes, especially since Allen played much less of a role at Microsoft after his 1982 bout with cancer and his growing disillusionment with Gates' confrontational leadership style.

Having just spent four years writing an exhaustive history of the world's leading atmospheric monitoring station, Hawaii's Mauna Loa Observatory, I disagree. Debates, arguments, and leadership flaws, whether in the low-pressure environment of a remote station at 11,200 feet or in the high-pressure environment of a startup company, are the sparks that illuminate the organization's history. Allen would have short-changed his readers had he failed to describe the disputes.

Allen even describes an expletive-laden temper salvo directed by Steve Jobs against a hapless Apple employee while he and Gates watched with surprise. Leadership antics like these will provide business analysts, academics, and, yes, psychologists much to ponder when they study the astonishing success of Microsoft and Apple.

Whether these disclosures have burnt the bridge in the four-decade relationship between Allen and Gates remains to be seen. In January 2011, four months before the release of *Idea Man*, Allen was in Albuquerque to dedicate Startup, a personal computer museum gallery, to the memory of Ed Roberts, who died in April 2010. When I asked Allen about his book, he said he was concerned how Gates would react.

Gates seems to have mellowed over the years. After he joined Allen in Albuquerque in the mid-'70s, the teenage-looking Gates sometimes had major battles with the burly Roberts, a former Air Force officer who expected respect. Last year, when Gates learned that Roberts was near death, he flew across the country to spend several hours with him days before Roberts died.

Allen writes in *Idea Man* that Gates regularly visited him in 2009 when he was hospitalized with his second battle with cancer: "He was everything you'd want from a friend, caring and concerned." Based on their history, it seems likely the two billionaires will eventually make their peace, perhaps while agreeing to disagree on some points. After all, many Microsoft customers who have a love-hate relationship with the company's software (including me) keep going back for more.

After Microsoft

Microsoft made its founders two of the world's richest men, and *Idea Man* follows Allen's account of the MITS-Microsoft years with highlights about his life, business, and philanthropy. He enthusiastically discusses his billionaire lifestyle, including his sports teams, his love affair with the guitar, and his far-flung travel adventures aboard his mega-yachts.

Much more important to us makers than the celebrity name-dropping and travel stories are the details of Allen's business successes and failures, his founding of the Allen Institute for Brain Science, and his carefully restored World War II-era aircraft collection.

Then there's Allen's partnership with Burt Rutan that culminated in SpaceShipOne, the first privately developed and launched reusable, manned spacecraft. The historic SpaceShipOne, which earned the $10 million Ansari X Prize, is now suspended between Lindbergh's Spirit of St. Louis and Chuck Yeager's Bell X-1 at the Smithsonian Air and Space Museum in Washington, DC.

Lessons for Makers

Idea Man provides important tips and lessons for today's generation of makers, some of whom might even now be developing what might become the next billion-dollar technology or product. Here are some lessons I've gleaned from its pages and between the lines.

Does your idea pass the balloon test? Good ideas and futuristic visions don't guarantee successful products and ventures. As Allen wrote about his pre-Altair days with Gates, "Each time I brought an idea to Bill, he would pop my balloon."

Texas-style handshake agreements with partners, supporters, and customers are great. I sold millions of books to RadioShack over handshakes and purchase orders. But Allen's experience suggests it's best to follow handshakes with carefully drafted agreements that all concerned are willing to sign.

Use care and prudence when working and dealing with partners and financial backers.

Get to know your partners and their idiosyncrasies before signing on with them.

Carefully read any agreement or contract before you sign it!

Partnerships are two-way arrangements. So get to know yourself. Are you living up to your agreements? Is your management style reasonable or do you create chaos?

A partnership agreement should provide contingencies for all eventualities. For example, the partners should agree to pursue arbitration in the event of a serious disagreement. The agreement should cover what happens should a partner be incapacitated or die.

Never, never, never release imperfect products! Delaying a promised new product is always better than releasing a defective one.

Treat your customers with the respect they deserve.

As Roberts learned so well, if your first products don't succeed, try again.

Thomas Jefferson:

Maker in Chief

There is no "Maker in Chief" in the U.S. Constitution, but at least one president, Thomas Jefferson, certainly deserves that title.

Jefferson is most famous for drafting the Declaration of Independence and serving as the third president of the United States. It's less well known that he was also an experienced architect, surveyor, locksmith, and amateur scientist. And he was an innovator who made improvements in the design of clocks, instruments, and the polygraph copying machines that duplicated his letters as he wrote them.

Visitors to Monticello, Jefferson's mountaintop home near Charlottesville, Virginia, can quickly become acquainted with Jefferson's maker side. Walking toward the front porch, you can see the entire house and its famous dome—all designed by Jefferson. "Architecture is my delight," he told a visitor, "and putting up and pulling down one of my favorite amusements."

FIGURE 28-1. The west side of Jefferson's Monticello home.

The Wind Vane

Another clue at Monticello is the large weather vane atop the front porch. The vane's shaft extends through the roof of the porch to the ceiling, where it is attached to a pointer that indicates the direction of the wind on a compass rose. Jefferson and his family could check the wind direction simply by looking through one of the nearby windows.

FIGURE 28-2. The wind vane over the front portico of Monticello.

The Great Clock

Above the front entrance is another Jefferson innovation, his double-sided Great Clock. The clock's movement and its dial indicating hours, minutes, and seconds are installed in a wood frame mounted inside the house, over the entrance door. A second dial that indicates the hour is mounted outside above the door.

The Great Clock was built to Jefferson's specifications by Philadelphia clockmaker Leslie & Price in 1793 and installed at Monticello in 1804. Its mechanism is powered by six cannonball-like weights suspended by a rope and pulley along the corner of the right side of the entrance hall. The days Sunday through Thursday are marked on the wall adjacent to the weights; as the weights descend, the topmost weight indicates the day of the week. Eventually, the weights drop through a circular hole in the floor to the basement, where Friday and Saturday are marked on the wall.

FIGURE 28-3. The outside dial of the Great Clock is over the entrance to the front portico.

A Chinese gong indicates each hour. The gong's striking mechanism is powered by a set of eight weights suspended by a rope and pulley along the corner of the left side of the entrance door.

Each Sunday, Jefferson climbed a folding ladder to wind the clock. When not in use, one side of the ladder was pushed up to merge it with the other side. The folding ladder was built in Monticello's wood shop, known as the joinery.

Jefferson's obsession with accurate timekeeping was closely related to his interest in astronomy and telescopes, several of which he purchased over the years. He described astronomy as "the most sublime of all the sciences."

Astronomy also served a practical purpose, for Jefferson was committed to measuring the geographic coordinates of Monticello as accurately as possible. With help from President James Madison—who lived nearby and often spent time at Monticello—Jefferson measured the timing of the annular solar eclipse of 1811. The times he and Madison measured were used by amateur eclipse expert William Lambert to calculate the longitude of Monticello as -78.50°, which is very close to Google Earth's -78.45°.

The Spherical Sundial

During Jefferson's time, the accuracy of clocks depended on a high-resolution sundial. Jefferson had probably seen spherical sundials in Europe, and he designed one for Monticello that was a 10½-inch sphere made from carefully inscribed locust wood fitted with a movable sundial blade. The spherical sundial was built in the joinery to his exact specifications. The fate of the original is unknown, but the Thomas Jefferson Foundation commissioned a replica based on Jefferson's detailed design. It is now installed at Monticello near where the original stood.

FIGURE 28-4. A replica of Jefferson's spherical sundial.

The Open-Source Plow

Perhaps the most down-to-earth of Jefferson's innovations was his design of an improved moldboard for plows. The moldboard is the section of the plow that lifts and turns the soil cut by the plow's leading edge. Jefferson claimed his moldboard was more efficient than previous designs.

He did not apply for patent protection for his design. Instead, he sent models of his moldboard design to others along with details about its design and construction.

Jefferson's Tools

While Jefferson was the architect of Monticello and the designer of many of its clocks, instruments, and mechanical contrivances, he did not personally build all of these things. But this does not disqualify him from being a hands-on maker. Jefferson's papers and correspondence mention his tools and that he took some of them to Paris when he served as U.S. minister to France from 1785 to 1789. Several of Jefferson's contemporaries wrote about his many personal tools, including Isaac Granger, who was born into slavery at Monticello and whom Jefferson took to Philadelphia, where he learned to be a tinsmith. When interviewed by the Rev. Charles Campbell in 1847, Granger recalled, "My Old Master was neat a hand as ever you see to make keys and locks and small chains, iron, and brass. He kept all kind of blacksmith and carpenter tools in a great case with shelves to it in his library."

Presidential Props

While addressing a dinner to honor Nobel Prize winners from the Western Hemisphere on April 23, 1962, President John Kennedy said, "I think this is the most extraordinary collection of talent, of human knowledge, that has ever been gathered together at the White House, with the possible exception of when Thomas Jefferson dined alone." If President Jefferson were alive today, we'd like to think he'd be among the many *Make:* magazine readers.

The Kit That Launched the Tech Revolution

29

Personal computers, laptops, and tablets were only a dream in 1975. Back then, electronics hobbyists were mesmerized by the January 1975 issue of *Popular Electronics* magazine (Figure 29-1). The cover showed a metal box with rows of toggle switches and LEDs under a label that read "Altair 8800." Boldly printed over the photo were these words: "Project Breakthrough! World's First Minicomputer Kit to Rival Commercial Models."

The Altair was developed by Micro Instrumentation and Telemetry Systems (MITS), a nearly bankrupt company in Albuquerque, New Mexico. The company's president and chief engineer was the late Ed Roberts (Figure 29-2), a no-nonsense visionary who had dreamed of building his own computer since high school.

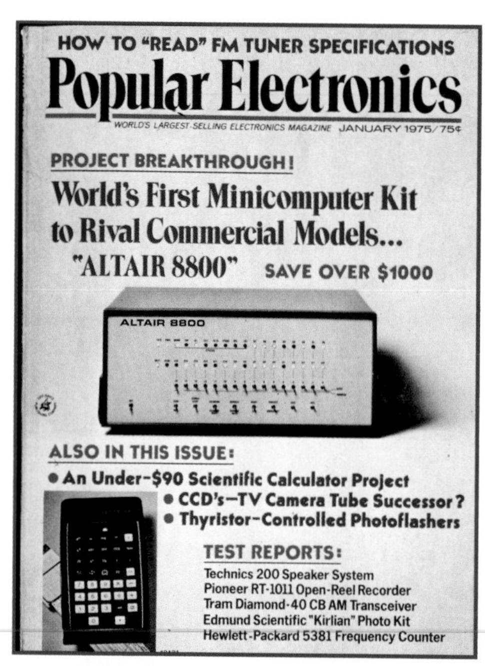

FIGURE 29-1. Though primitive by today's standards, the Altair 8800 is widely credited with jump-starting the personal computer era.

FIGURE 29-2. Ed Roberts, president and chief engineer of MITS. I took this photo in Ed's office about 20 feet from where Bill Gates would end up a few years later.

MITS and *Popular Electronics* come to mind every time I browse *Make:* magazine. Will something as revolutionary as the Altair 8800 someday emerge from these pages? Perhaps the Altair story will inspire you to transform your dream into the next big thing.

The MITS Story

MITS was founded in 1969, when Robert Zaller, Stan Cagle, and I met at Ed Roberts' house in Albuquerque (Figure 29-3) to form a company to manufacture a model rocket light flasher that I had described in the September 1969 issue of *Model Rocketry*

FIGURE 29-3. Alburqueque, NM—the place where it all began.

magazine. That article began my career as a writer, but Ed had much bigger ambitions.

After we sold only a few hundred rocketry gadgets, we decided to try something new during the summer of 1970. That spring I had written my first article for *Popular Electronics*, a feature on LEDs. When I asked if they'd like an LED lightwave communication project and kit to accompany the feature story, the answer was yes. Ed and Bob designed a prototype we called the Opticom (Figure 29-4) that could send voice up to 1,000 feet. *Popular Electronics* published both articles in November 1970.

FIGURE 29-4. The author's model rocket light flasher shown here launched the forming of MITS, Inc. in 196.

I soon left MITS to become a full-time writer. Ed stayed on to develop calculators. His article about the MITS 816, the first digital calculator kit, also made the cover of *Popular Electronics*. All went well until serious competition arrived from Japan. MITS was nearly bankrupt by 1974.

That summer MITS' *Popular Electronics* connection would come to the rescue. The cover story of the July 1974 *Radio-Electronics* magazine, *Popular Electronics'* main competitor, was a breakthrough project: the Mark-8, a DIY microcomputer designed by Jonathan Titus around Intel's 8008 7-bit microprocessor. The article offered a manual and a printed circuit board but not a complete kit. Nevertheless, the Mark-8 lit the rivalry fuse.

The Altair 8800

Popular Electronics editor Art Salsberg and technical editor Les Solomon knew that Ed Roberts and MITS engineer Bill Yates were working on a microcomputer project using a more advanced processor, Intel's new 8080 chip. They agreed to publish a major cover story. One evening Ed called to ask if I would stop by to see the first prototype, so I hopped on my bicycle and rode the five blocks to MITS. Ed and Bill were standing by a metal box about the size of a thick briefcase on a workbench. Its front panel was lined with rows of switches and LEDs. Hanging from the wall were Bill's intricate layout patterns for the PC boards inside the box.

Ed invited me to take a close look, then asked: "How many do you think we'll sell?" Based on sales of MITS model rocket gear, the Opticom, and the calculators, I was not optimistic that a bare-bones computer would do much better. So I said a few hundred at most.

Ed was disappointed by my response, for he was confident the computer would easily sell many hundreds. But we were both wrong. In the months following the Altair 8800 article in *Popular Electronics*, MITS sold thousands of assembled and kit Altairs, even though the early models had only 256 bytes of RAM and no keyboard or monitor beyond their front panel switches and LEDs. The price of the basic kit was $439, around $1,925 in today's dollars.

DIY Altairs and Other Vintage Computers

The Altair 8800 lives on in pampered working versions cared for by enthusiastic computer historians, engineers, and hobbyists. You can share their passion for the earliest days of personal computing with replica Altair kits, PC boards, and assembled versions available online.

Classic computer collector Rich Cini designs replica PC boards of early computers, including the Altair 8800. He has also developed an Altair emulator that programmers will find interesting (http://classiccmp.org/altair32/). Cini highly recommends http://s100computers.com/ and the N8VEM Home Brew Computer Project (http://obsolescence.wix.com/obsolescence#!n8vem-overview/csbv). These sites specialize in PC boards compatible with the S-100 bus that Ed Roberts designed to interconnect the boards of the original Altair 8800.

Grant Stockly (http://altairkit.com/) and Mike Douglas (http://altairclone.com/) sell Altair replica kits complete with custom-made cabinets carefully copied from the original Optima housing. Douglas' price is notable in that it's identical to the original MITS price—$439—even though a dollar in 1975 equals around $4.50 today.

The Altair's Legacy

Computers require a language and programs. Paul Allen knew that very well when he spotted the Altair on the cover of *Popular Electronics* at Out of Town News, a Harvard Square newsstand. He bought the magazine and hurried to the Harvard dormitory where his friend Bill Gates (Figure 29-5) resided. Allen and Gates soon contacted Ed Roberts, and the collaboration that followed resulted in the founding of Microsoft. MITS was their first customer.

Ed's Altair and Microsoft's version of the BASIC programming language ignited the revolution that soon led to the personal computers introduced by Apple, RadioShack, IBM, and a host of other firms. The Altair's role is not forgotten. The one Ed gave me for writing the Altair manual is now in the Smithsonian.

Learn More

"The Altair Story: Early Days at MITS," by Forrest M. Mims III, *Creative Computing*, November 1984, http://atarimagazines.com/creative/v10n11/17_The_Altair_story_early_d.php.

Idea Man by Paul Allen, reviewed in *Make:* Volume 27, http://makezine.com/2011/05/10/startups-origins-of-the-pc-revolution-by-forrest-m-mims-iii/

StartUp Gallery (Figure 29-6) at the New Mexico Museum of Natural History and Science in Albuquerque, http://startup.nmnaturalhistory.org/.

When Projects Fail

30

My DIY wearable gadget never sold—but it led to my first article, a kit company, and a career in electronics and science.

Projects that fail are rarely published—but they teach valuable lessons that often lead to success. If you've spent much time designing and building projects, you know this well. I certainly do. Some of my failed projects made a major impact on my career in electronics and science.

During my senior year at Texas A&M University, in 1966, Texas Instruments announced the development of a powerful LED that emitted several milliwatts of invisible near-infrared, about the same power output as a small flashlight. My great-grandfather had been totally blinded by a dynamite explosion when he was a young man, and the new LED gave me an idea for building a travel aid for the blind.

So I hitchhiked to Dallas to meet Edward Bonin, one of TI's LED engineers. The new LEDs cost $356 each, about $2,671 in today's money. Dr. Bonin said he would give me, a rank amateur, one of the LEDs if I could build a

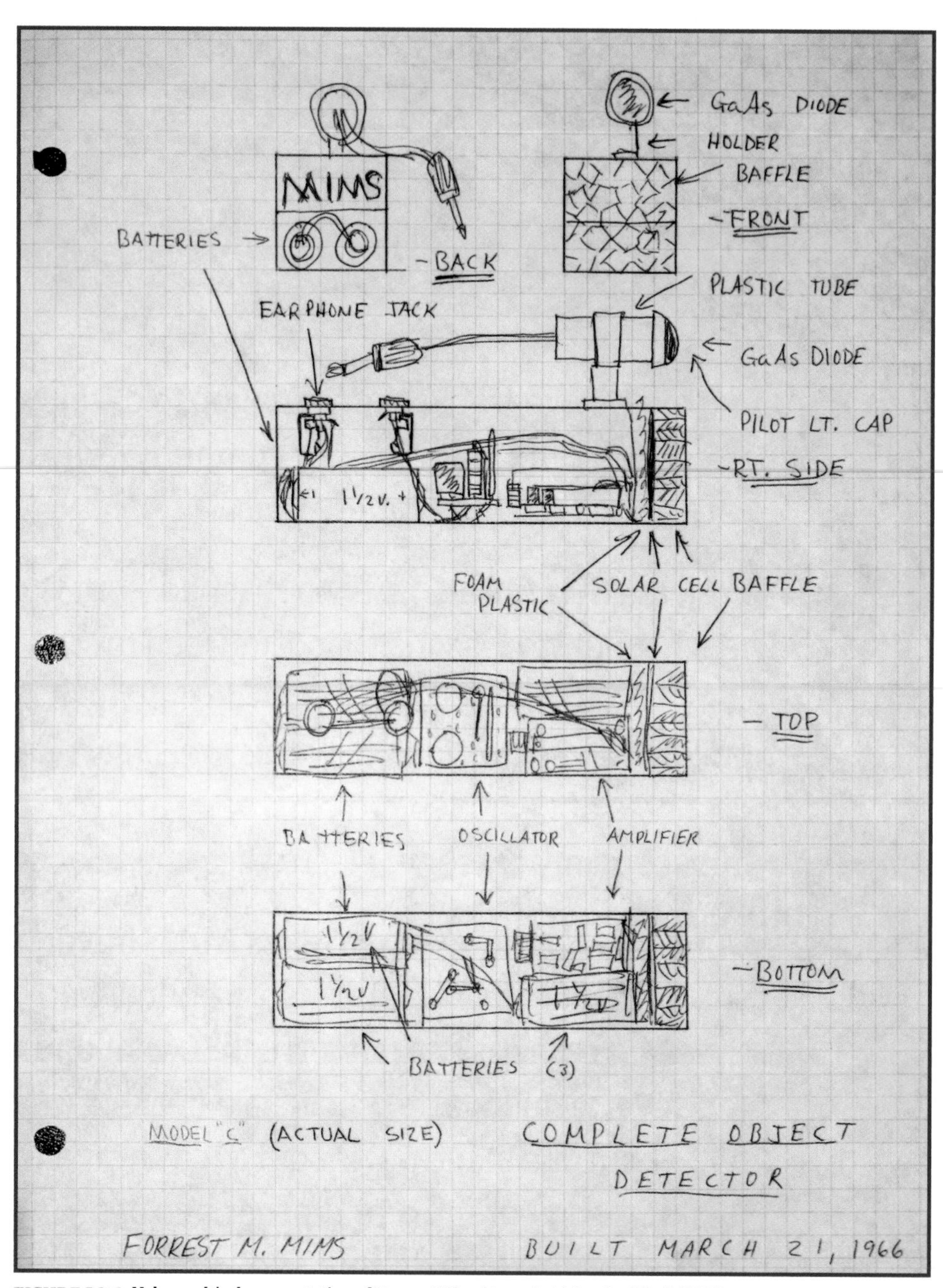

FIGURE 30-1. Holographic documentation of Forrest Mims' travel aid for the blind, 1966.

circuit that would generate the pulses needed to make the travel aid. I modified a 1-transistor Morse code practice oscillator board sold by a radio and TV repair shop for 99 cents and sent it to Bonin. He approved the circuit and sent it back, together with three of the sophisticated LEDs.

I quickly built and documented a prototype (Figure 30-2) and within a few days built a working travel aid that measured 2"×2"×4".

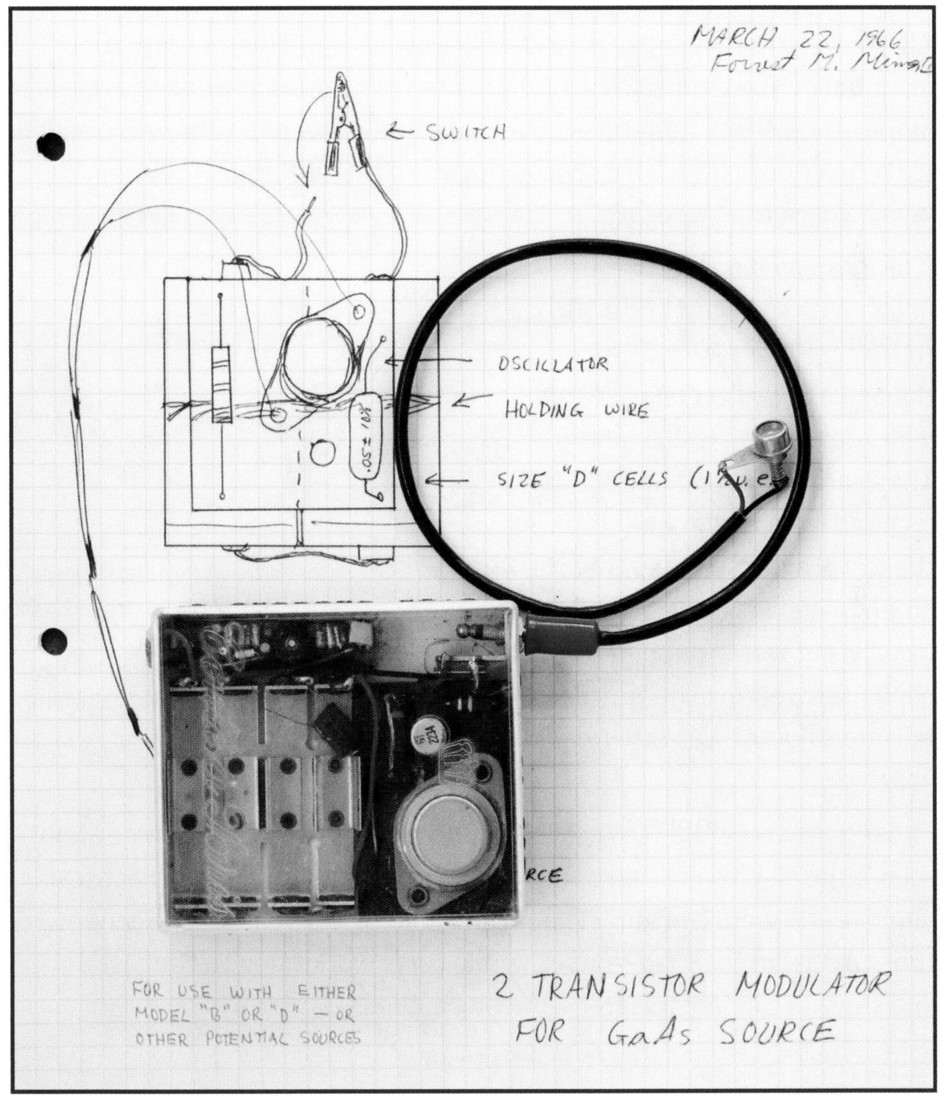

FIGURE 30-2. Prototype of Forrest Mims' travel aid for the blind, 1966.

Flashes of invisible infrared emitted by the aid were reflected by objects up to 10 feet away. The reflected IR was detected by a silicon solar cell, and the resultant photocurrent was amplified by a transistor amplifier (salvaged from a hearing aid) and sent to an earphone, which emitted a tone. The closer the object, the louder the tone.

I tested the travel aid with more than 20 blind children and adults. It worked well, but the need to hold it in one hand was a drawback. Eventually I assembled the entire device on a pair of sunglasses (Figure 30-3). All the electronics were installed inside two 3/8"-diameter brass tubes mounted on the temples, the LED transmitter in one tube and the receiver in the other. A tiny hearing aid earphone in the receiver tube was coupled to the user's ear through a short length of plastic tubing.

The eyeglass aid worked well. It also received an Industrial Research 100 Award and a 1987 runner-up Rolex Award. But in the end, the project I had spent years developing was a failure. The hearing aid companies I approached about manufacturing the travel aid responded that the potential liability was much too risky. What would happen if a blind user wearing the travel aid fell into a hole or was otherwise injured?

FIGURE 30-3. All the electronics were installed inside two 3/8"-diameter brass tubes mounted on the temples, the LED transmitter in one tube and the receiver in the other. A tiny hearing aid earphone in the receiver tube was coupled to the user's ear through a short length of plastic tubing.

Though the travel aid was never manufactured, it taught me more about solid-state electronics and optics than my friends majoring in electrical engineering were learning. They were building old-fashioned vacuum tube circuits in their lab courses, while I was working with transistors and state-of-the-art infrared-emitting diodes.

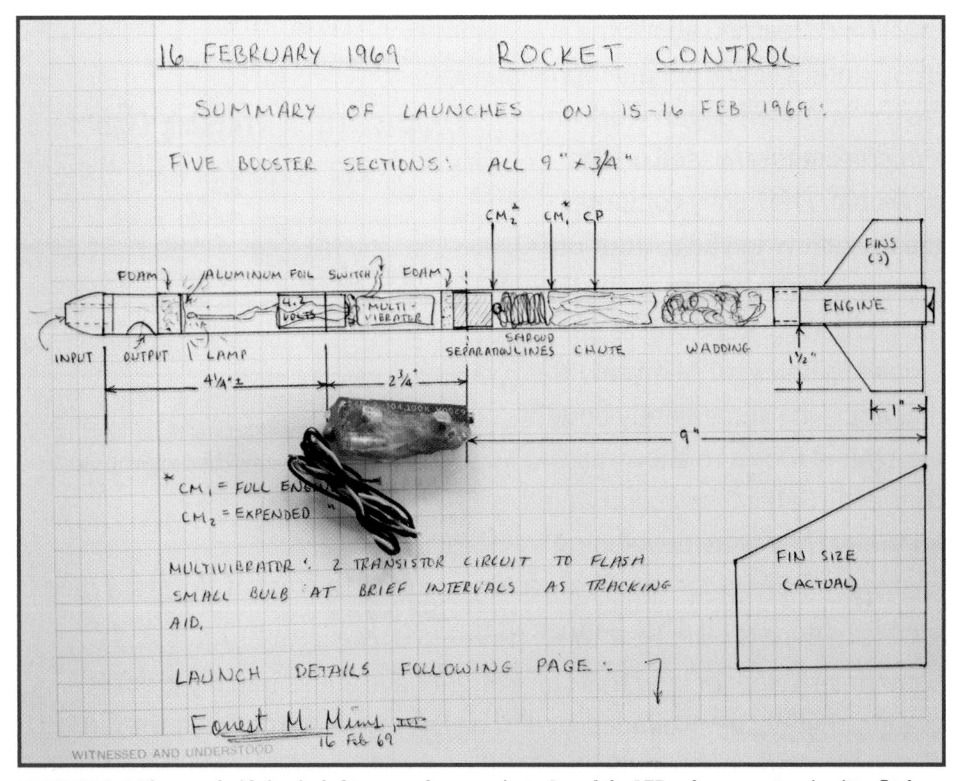

FIGURE 30-4. The travel-aid circuits led to several new projects. I used the LED pulse generator circuit to flash a tracking light in night-launched model rockets I was flying to test a new kind of guidance mechanism.

Ed Roberts and I were then assigned to the Laser Division of the Air Force Weapons Laboratory. We often talked about selling electronics kits through magazines like *Popular Electronics* and *Radio-Electronics*. When the light flasher article was published, we decided to form a company to build and sell light flashers and other model rocketry gear. We called it Micro Instrumentation and Telemetry Systems (MITS).

I eventually left MITS to pursue a new career as an electronics writer. Ed stayed and introduced a string of new products. I wrote the instruction manuals for some of them. I also introduced Ed to Leslie Solomon, the technical editor of *Popular Electronics*.

In 1974, Ed learned about the 8080, Intel's new 8-bit microprocessor, and soon began work on a microcomputer based on the new chip. The hobby computer era took off when Ed's Altair 8800 appeared on the cover of the January 1975 issue of *Popular Electronics*. When Paul Allen saw the magazine, he immediately bought a copy and took it to show his friend Bill Gates. They soon called Ed to say they were developing a version of BASIC for the Altair, and you know the rest of the story (see Chapter 29, "The Kit That Launched the Tech Revolution").

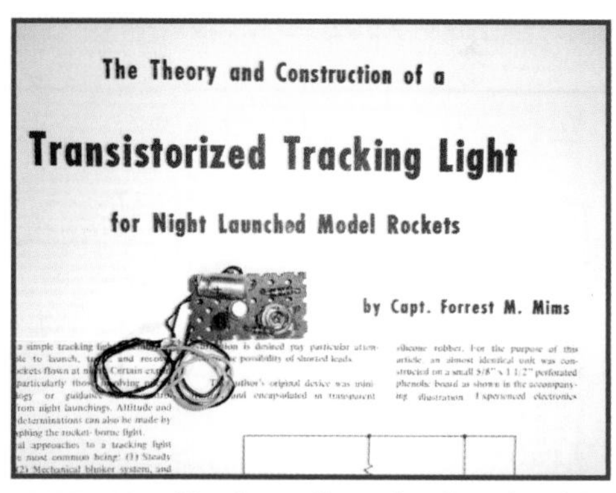
The Theory and Construction of a

Transistorized Tracking Light

for Night Launched Model Rockets

by Capt. Forrest M. Mims

FIGURE 30-5. After George Flynn, the editor of *Model Rocketry* magazine, watched one of those flights, he asked me to write an article about the light flasher. It was published in September 1969.

Sometimes I wonder how this story might have ended had my great-grandfather not been blinded, or if TI hadn't invented the first infrared LED. Of course it's impossible to know in advance what might come from a failed or abandoned project—and that's motivation enough to press ahead.

Going Further

Have you developed a project that failed for technical or other reasons? Think about how it might have advanced your knowledge, and tell us at http://makezine.com/2015/02/25/when-projects-fail/.

And consider beginning a new project that's got only a marginal chance for success. My view is that every project is like a course in tech school or college, for the spinoffs from a failed project are sometimes as significant as those from the great successes.

Index

Numbers

A

B

C

D

E

F

G

H

I

J

K

L

M

N

O

P

Q

R

S

T

X

Y

Z